# Topological quantum field theories from subfactors

# CHAPMAN & HALL/CRC

## Research Notes in Mathematics Series



**Submission of proposals for consideration**

Suggestions for publication, in the form of outlines and representative samples, are invited by the Editorial Board for assessment. Intending authors should approach one of the main editors or another member of the Editorial Board, citing the relevant AMS subject classifications. Alternatively, outlines may be sent directly to the publisher's offices. Refereeing is by members of the board and other mathematical authorities in the topic concerned, throughout the world.

**Preparation of accepted manuscripts**

On acceptance of a proposal, the publisher will supply full instructions for the preparation of manuscripts in a form suitable for direct photo-lithographic reproduction. Specially printed grid sheets can be provided. Word processor output, subject to the publisher's approval, is also acceptable.

Illustrations should be prepared by the authors, ready for direct reproduction without further improvement. The use of hand-drawn symbols should be avoided wherever possible, in order to obtain maximum clarity of the text.

The publisher will be pleased to give guidance necessary during the preparation of a typescript and will be happy to answer any queries.

**Important note**

In order to avoid later retyping, intending authors are strongly urged not to begin final preparation of a typescript before receiving the publisher's guidelines. In this way we hope to preserve the uniform appearance of the series.


**CRC Press UK**

Chapman & Hall/CRC Statistics and Mathematics
Pocock House
235 Southwark Bridge Road
London SE1 6LY
Tel: 020 7450 7335

Vijay Kodiyalam and V S Sunder

# Topological quantum field theories from subfactors

CRC Press
Taylor & Francis Group
Boca Raton London New York

CRC Press is an imprint of the
Taylor & Francis Group, an **informa** business
AN A K PETERS BOOK

CRC Press
Taylor & Francis Group
6000 Broken Sound Parkway NW, Suite 300
Boca Raton, FL 33487-2742

First issued in hardback 2019



ISBN-13: 978-1-58488-241-1 (pbk)
ISBN-13: 978-1-138-44210-8 (hbk)



**Visit the Taylor & Francis Web site at**
**http://www.taylorandfrancis.com**

**and the CRC Press Web site at**
**http://www.crcpress.com**

**Library of Congress Cataloging-in-Publication Data**

Catalog record is available from the Library of Congress.

*To*

*Maitreyi and Radha*

# Contents

# Preface

These notes are the contents of a course of lectures given by the second author at the Institute of Mathematical Sciences, Chennai. The aim of the course is probably best described by the following 'blurb' which accompanied the announcement of this course:

*The course will begin with axioms for a TQFT, then go through the Turaev-Viro prescription for constructing such a TQFT, and finally work through Ocneanu's method of starting with a 'finite depth hyperfinite subfactor' and obtaining the kind of data needed to invoke the Turaev-Viro machine.*

*An attempt will be made to make this as self-contained as possible. In particular, no background on subfactors will be assumed; the idea will be to develop the necessary facts from the theory of subfactors as and when they are needed.*

The audience for these lectures was a mix of beginning doctoral students and working research mathematicians of various specialisations (topology, algebraic geometry, commutative algebra). So, the various ingredients that go into Ocneanu's TQFTs - which are obtained by adapting the Turaev-Viro recipe to the context of bimodules of finite type over $II_1$ factors - had to each be introduced and discussed. The exposition was midway between working out all details of everything (which would have taken far too much time) and merely stating the facts without proofs (which would have been too mysterious and not too satisfying).

These notes should be accessible to second year graduate students, who have had a minimal exposure to Hilbert space theory and some topology, and are blessed with a modicum of mathematical maturity. In particular, working mathematicians (of any specialisation) and mathematical physicists should not face any serious difficulties with this text.

No claim is made to any originality in these notes; all of the material contained here can be found elsewhere in the literature, in some form or the other. The primary source for much of the material was the recent book by Evans and Kawahigashi

([EK] in the references here); however, several more details have been worked out here than there; also, several proofs here are different from the ones given there. The only reason for even attempting to write up these notes is that perhaps they would be sufficiently complete and accessible to a wider class of people than the Evans-Kawahigashi book referred to above.

The text is far from being meticulous about giving references; since these are primarily intended as 'lecture notes', only a small number of references are included, and are merely meant to point the reader to other places with more numerous references to the theory. However, a brief section on 'Bibliographical Remarks' is included at the end, for the reader's sake. Also included for the reader's convenience is an index of notation and terms used in the text.

It is a pleasure to record our gratitude to all the people who patiently sat through the entire duration of the course mentioned above and gave the necessary momentum for carrying out this work. We would also like to thank Yasu Kawahigashi for having been unfailingly prompt, efficient and helpful, whenever we had any doubts or questions (of a technical or bibliographical nature). Finally, this is the place to acknowledge the second author's gratitude to Fred Goodman for: (a) sweating with him through some of the details presented here, (b) having been the perfect host during the Winter semester of 1999 when (a) was carried out, and (c) being a great guy.

# Chapter 1

# Preliminaries on TQFTs

## 1.1 Introduction

One of the hopes that motivated the writing up of these notes was that it might be accessible to an audience of which very little specific background preparation was demanded; the hope was that it might be possible to address a person with nothing more than a 'modicum of mathematical maturity' and to describe (to such a person) some of the beautiful concepts and ideas that go into this construction (by Ocneanu) of a 'topological quantum field theory from a finite-depth subfactor'. While the necessary background from the theory of von Neumann algebras is presented (quite rigorously and precisely, and essentially 'from scratch') in the third chapter, a prospective reader with no background in topology is offered a somewhat informal discussion, in the Appendix, on some of the basic terms used. After a very non-technical description of oriented manifolds in §A.1, we proceed in §A.2 to a brief discussion of simplicial complexes, where we also include a short discussion on the 'edge-path groupoid' of a simplicial complex.

In §1.1, we proceed to a more formal and elaborate discussion of 'quantisation functors on cobordism categories' which may be thought of as precursors to 'topological quantum field theories'. We spell things out in some detail here, if at least to highlight the problem - see [Tur] for a discussion of this question and some of its consequences - of determining whether a given quantisation functor (such as the TQFT we will be concerned with in these notes) has the property of being 'cobordism generated'. The next section is devoted to gathering together the definition and some basic facts concerning (finite-dimensional) fusion algebras, their dimension functions and the notion of such a fusion algebra 'admitting intertwiner spaces'.

Chapter 2 is devoted to the construction of the invariant and the verification that the given prescription does indeed define a TQFT. The strategy is as follows: in §2.1, we start with the 'input' of a 'fusion algebra which admits intertwiner spaces' in the sense described in §1.2; we describe what it means to have a 'quantum $6j$ symbol' subordinate to the above 'input'; loosely speaking, it is a procedure of assigning a number to a 'state on a tetrahedron' - this being a manner of (i) choosing a total order on the set of vertices of a tetrahedron, (ii) labelling each vertex of the tetrahedron by an element from the distinguished basis of the fusion algebra, and (iii) labelling each face of the tetrahedron by an element of a distinguished orthonormal basis from the 'intertwiner Hilbert space' associated to that face - where the assignment is required to satisfy some 'compatibility constraints'; and we conclude §2.1 by showing how, given a triangulation $\Delta$ (i.e., simplicial decomposition) of a closed oriented 3-manifold $M$, and a total order '$\leq$' on the set of vertices of $\Delta$, as well as a choice of a system $\mathcal{B}$ of orthonormal bases for various Hilbert spaces that get specified in the process, we may obtain a complex number $\langle(M, \Delta, \leq, \mathcal{B})\rangle$. (We should mention that the device of choosing the appendage of a total order is not something that will be found in [EK]; our reason for this device is the belief that it helps clarify various definitions and arguments. We should perhaps also mention that our proof of 'invariance' differs from the one found in [EK], in that we restrict ourselves to only *bona fide* triangulations; and the reader without topological intuition will not be called on to perform the mental calisthenics of visualising 'a triangulation in which two tetrahedra share three faces in common'.) We then take up, in §2.2, the task of showing that the number $\langle(M, \Delta, \leq, \mathcal{B})\rangle$ is independent of the choices of the auxiliary system $\mathcal{B}$ of bases, the total order '$\leq$', and finally the triangulation $\Delta$, and is hence indeed an 'invariant' of the manifold $M$; and we finally show how these arguments need to be modified in order to handle manifolds with boundary in order to obtain a (2+1)-dimensional TQFT.

The whole aim of these notes is to show how the structure of a so-called 'finite-depth subfactor' lends itself most naturally to the construction of a TQFT according to the prescription outlined in chapter 2. Chapter 3 is devoted to a discussion of the necessary background from the theory of von Neumann algebras in general, and the theory of 'bimodules of finite type' over $II_1$ factors in particular. With the exception of the unproved facts which are listed in §3.1, the treatment here is reasonably self-contained. In particular, even the reader who only wishes to learn the basics of subfactor theory would find the treatment here quite short but nevertheless fairly detailed on such matters as '$M$-valued inner products', contragredients and tensor-products (or 'Connes' fusion' as it is sometimes called) of bimodules. After reviewing all these matters in the first four sections, we finally discuss, in §3.5, the small amount of subfactor theory that is necessary to reach the goal announced in the first sentence of this paragraph.

In chapter 4, we work out some details of the TQFT associated to one of the simplest 'finite-depth subfactors' - the one given by the 'crossed-product of a $II_1$ factor by the outer action of a finite group $G$'. There is really no von Neumann algebra theory at all in evidence here, only fairly simple group theory. We see that the TQFT, in this case, is closely related to group-homomorphisms into $G$ from the fundamental group(oid) of a closed manifold. This is borne out by our computation of the numerical invariant of a closed 3-fold, as well as of the dimensions of the Hilbert spaces associated to the oriented surfaces of different genera. Finally, in §4.2, we briefly discuss the quantum double of a finite group and its representation theory, and show how our computations, in §4.1, of the dimensions of the Hilbert spaces associated to surfaces of various genera, tally with the numbers predicted by Verlinde's formula, when applied to the representation theory of the quantum double.

## 1.2 Quantisation functors

We shall be dealing with smooth oriented compact manifolds (and we shall deal only with such manifolds). For such a manifold $M$, we shall write $M^*$ for the same manifold, but with the reversed orientation. Suppose that the boundary of a $(d+1)$-dimensional manifold $M$ decomposes as $\partial M = \Sigma_1^* \coprod \Sigma_2$ - where $\coprod$ denotes disjoint union; then $M$ is called a **cobordism** from $\Sigma_1$ to $\Sigma_2$ and we write $M \in Cob(\Sigma_1, \Sigma_2)$. (Note that the same $\partial M$ may admit several such decompositions; for instance, $M \in Cob(\emptyset, \partial M)$ as well as $M \in Cob((\partial M)^*, \emptyset)$; in general if $\partial M$ has $n$ components, there will be $2^n$ ways of regarding $M$ as a cobordism.)

Cobordisms $M, N \in Cob(\Sigma_1, \Sigma_2)$ are said to be **equivalent** if there exists an orientation preserving diffeomorphism of $M$ onto $N$ which restricts to the identity on the boundary. We shall write $[M]$ for the equivalence class of a cobordism $M$.

Two cobordisms 'with matching attributes' can be *glued together*: i.e., if $M \in Cob(\Sigma_1, \Sigma_2)$ and $N \in Cob(\Sigma_2, \Sigma_3)$, write

$$[N] \circ [M] \; = \; [M \cup_{\Sigma_2} N] \; . \tag{1.2.1}$$

(It should be checked that the right side of equation (1.2.1) depends only on the classes $[M], [N]$.)

Finally, we may construct a category $\mathcal{C}_0$ as follows: the objects of $\mathcal{C}_0$ are smooth oriented $d$-dimensional manifolds, whose typical member will be usually denoted by $\Sigma$ or a suitably decorated version thereof; the morphisms of $\mathcal{C}_0$ are equivalence classes of cobordisms - thus,

$$Hom(\Sigma_1, \Sigma_2) \; = \; \{[M] \; : \; M \in Cob(\Sigma_1, \Sigma_2)\} \; ,$$

and composition of morphisms is as in equation (1.2.1). It is easily verified that the above definitions do indeed endow the structure of a category on $\mathcal{C}_0$, where $[\; \Sigma \times [0,1] \;]$ plays the role of the identity morphism on $\Sigma$. (It should be noted that considering equivalence classes of cobordisms, rather than just cobordisms, is necessitated by the identity axiom as well as by the axiom of associativity of composition of morphisms in a category.)

The following notions make sense in the category $\mathcal{C}_0$ of the last paragraph: (a) an empty object $\emptyset$, (b) disjoint unions, (c) an involution, denoted

$$\begin{array}{rcl} Obj(\mathcal{C}_0) \ni \Sigma & \mapsto & \Sigma^* \in Obj(\mathcal{C}_0) \\ Hom(\Sigma_1, \Sigma_2) \ni [M] & \mapsto & [M^*] \in Hom(\Sigma_2, \Sigma_1) \ , \end{array} \quad (1.2.2)$$

and (d) boundary, which is the map from morphisms to objects defined by $Hom(\Sigma_1, \Sigma_2) \ni M \mapsto \partial M = (\Sigma_1^* \coprod \Sigma_2)$.

Let us loosely call a category as above a **cobordism category**. We will not make a precise definition; in practice, we shall typically work with categories whose objects are 'suitably *marked* elements' of the above $\mathcal{C}_0$, and morphisms are also 'suitably marked cobordisms'. The specific situation will make clear how the above four notions should be interpreted. Also, when it is necessary to call attention to the underlying dimension, we shall refer to such a $\mathcal{C}$ as a $(d+1)$-dimensional cobordism category.

We shall be concerned with functors $V$ from a cobordism category into the category $k - mod$ of $k$-modules, where $k$ will always denote a commutative ring, which comes equipped with identity 1 and an involution (= automorphism of order 2) $\lambda \mapsto \bar{\lambda}$. (This level of generalisation is just for the duration of this section; from the next section onwards, we will only consider $k = \mathbb{C}$ (with conjugation as involution).) A **quantisation functor** $V$ on a cobordism category is a functor as above which is required to satisfy two requirements, which we shall label by (Q1) and (Q2).

**(Q1)** $$V(\emptyset) = k.$$

We need to set up some notation before we can state (Q2). So suppose $V : \mathcal{C} \to k - mod$ is a functor which satisfies (Q1). For $M \in Hom(\Sigma_1, \Sigma_2)$, we shall write $Z_M$ (rather than the more natural $V(M)$) for the $k$-homomorphism $Z_M : V(\Sigma_1) \to V(\Sigma_2)$. (The reason for this notation is twofold: first, this seems to be historically accepted notation; second, this will help to avoid possible subsequent confusion: for instance, the symbol $\emptyset$ might denote an object of $\mathcal{C}$ as well as a morphism in $Hom(\emptyset, \emptyset)$.)

Now if $M \in Hom(\emptyset, \Sigma)$ (which, in case $\mathcal{C} = \mathcal{C}_0$, is just the isomorphism class, rel boundary, of a $(d+1)$-dimensional manifold), define

$$\zeta_M = Z_M(1) , \tag{1.2.3}$$

which makes sense since $Z_M : V(\emptyset) \to V(\Sigma)$ and $1 \in k = V(\emptyset)$ (by (Q1)).

If $\Sigma = \emptyset$ above, then $\zeta_M \in k$; in this case, we write

$$< M >_V = \zeta_M . \tag{1.2.4}$$

Note that in case $\mathcal{C} = \mathcal{C}_0$, and $M$ is a $(d+1)$-dimensional manifold, then $< [M] >_V$ depends only on the isomorphism class $[M]$ of the manifold $M$. We are now ready to state the condition (Q2).

**(Q2)** There exists a hermitian sesquilinear* form $\langle \, , \, \rangle_\Sigma$ on $V(\Sigma)$ such that

$$\langle \zeta_{M_1}, \zeta_{M_2} \rangle_\Sigma = < (M_2)^* \circ M_1 > , \tag{1.2.5}$$

for every $M_1, M_2 \in Hom(\emptyset, \Sigma)$ .

LEMMA 1.2.1 *Suppose $V : \mathcal{C} \to k-mod$ is a quantisation functor. Then,*

*(1) If $\Sigma \in Obj(\mathcal{C})$, then $id_\Sigma \in Hom(\Sigma, \Sigma)$, and $Z_{id_\Sigma} = id_{V(\Sigma)}$.*

*(2) If $M \in Hom(\Sigma_1, \Sigma_2)$, then*

$$Z_M(\zeta_{M_1}) = \zeta_{M \circ M_1} , \ \forall \ M_1 \in Hom(\emptyset, \Sigma_1) , \tag{1.2.6}$$

*from which it follows that*

$$\langle Z_M(\zeta_{M_1}), \zeta_{M_2} \rangle_{\Sigma_2} = < M_2^* \circ M \circ M_1 >_V , \tag{1.2.7}$$

*for every $M_1 \in Hom(\emptyset, \Sigma_1)$, $M_2 \in Hom(\emptyset, \Sigma_2)$ .*

*(3) $Z_{id_\emptyset} = id_k$. Further,*

$$< id_\emptyset >_V = 1 = \langle 1, 1 \rangle_\emptyset ,$$

*and more generally,*

$$\langle \alpha, \beta \rangle_\emptyset = \alpha \bar{\beta} , \ \forall \ \alpha, \beta \in k .$$

*This is the requirement that $\langle \cdot, \cdot \rangle_\Sigma$ is linear in the first variable and 'conjugate-linear' in the second.

**Proof:** Both (1) and equation (1.2.6) in (2) are immediate consequences of the requirements of a functor; thus while (1) is a consequence of the requirement that a functor respect the identity morphism, the requirement that a functor respect composition of morphisms implies that

$$Z_M(\zeta_{M_1}) = Z_M(Z_{M_1}(1)) = Z_{M \circ M_1}(1) = \zeta_{M \circ M_1} .$$

Equation (1.2.7) is a consequence of equation (1.2.6) and equation (1.2.5).

The first equation in (3) is a special case of (1), while

$$< id_\emptyset >_V = Z_{id_\emptyset}(1) = id_k(1) = 1 .$$

Hence, by equation (1.2.5) - with $M_1 = M_2 = \emptyset$ - it follows that

$$\langle 1, 1 \rangle_\emptyset = \langle \zeta_\emptyset, \zeta_\emptyset \rangle_\emptyset = < id_\emptyset \circ id_\emptyset >_V = < id_\emptyset >_V = 1 .$$

The final assertion follows now from the definition of a sesquilinear form.

□

PROPOSITION 1.2.2 *Suppose $V : \mathcal{C} \to k-mod$ is a quantisation functor. Then,*

*(1)* $< id_\emptyset >_V = 1$ ;

*(2)* $M \in Hom(\Sigma_1, \Sigma_2) \Rightarrow M^* \in Hom(\Sigma_2, \Sigma_1)$, *and, in particular, if* $M \in Hom(\emptyset, \emptyset)$, *then* $< M^* >_V = \overline{< M >_V}$ ;

*(3)* $< M_1 \circ M_2 >_V = < M_1 >_V < M_2 >_V$ , $\forall M_1, M_2 \in Mor(\emptyset, \emptyset)$.

**Proof:** (1) See Lemma 1.2.1 (3).

(2) For the first implication, see (1.2.2). Notice, by (Q2), that if $M_1, M_2 \in Hom(\emptyset, \Sigma)$, then $\langle \zeta_{M_1}, \zeta_{M_2} \rangle_\Sigma = < M_2^* \circ M_1 >_V$. In particular, if $\Sigma = \emptyset$ (so that the 'cobordisms are closed'), this reads (thanks to the last assertion of Lemma 1.2.1 (3), and (Q2)):

$$< M_1 >_V \overline{< M_2 >_V} = \langle < M_1 >, < M_2 > \rangle_k = < M_2^* \circ M_1 >_V .$$

Set $M_1 = id_\emptyset$ to find that $< M^* >_V = \overline{< M >_V}$ for any closed morphism $M$, thereby establishing (2).

(3) follows by using (2) in conjunction with the equation displayed in the proof of (2).

□

DEFINITION 1.2.3 *A quantisation functor* $(V, Z, \zeta)$ *which satisfies the following condition, for all* $\Sigma$*, is said to be* **cobordism-generated**:

**(CG)** $$V(\Sigma) = k - span \, \{\zeta_M : M \in Hom(\emptyset, \Sigma) \} .$$

*In the sequel, we shall use the acronym* c.g.q.f. *to signify a cobordism-generated quantisation functor.*

The reason for the above definition lies in the following 'converse' to Proposition 1.2.2.

PROPOSITION 1.2.4 *If* $C$ *is a cobordism category, and if* $<>$ *is a* $k$*-valued invariant defined on the morphisms in* $Hom(\emptyset, \emptyset)$*, which satisfies properties (1), (2) and (3) of Proposition 1.2.2, then there exists a unique (up to natural isomorphism) c.g.q.f.* $V$ *defined on* $C$ *such that* $<> = <>_V$ .

**Proof:** For each $\Sigma \in Obj(C)$, let $\tilde{V}(\Sigma)$ denote the free $k$-module with basis $\{[M] : M \in Hom(\emptyset, \Sigma)\}$. Then the equation

$$\langle \sum_M \alpha_M [M], \sum_N \beta_N [N] \rangle_\Sigma = \sum_{M,N} \alpha_M \overline{\beta_N} < N^* \circ M > \quad (1.2.8)$$

is seen to define a sesquilinear form on $\tilde{V}(\Sigma)$, which is 'hermitian', since, for arbitrary $M, N \in Hom(\emptyset, \Sigma)$, we have:

$$\begin{aligned} \langle [N], [M] \rangle_\Sigma &= < M^* \circ N > \\ &= < (N^* \circ M)^* > \\ &= \overline{< N^* \circ M >} \text{ (by the assumed property (2))} \\ &= \overline{\langle [M], [N] \rangle_\Sigma} . \end{aligned}$$

Let $\mathcal{N}(\Sigma)$ denote the 'radical' of the form $< \cdot, \cdot >_\Sigma$, which is the set defined by $\mathcal{N}(\Sigma) = \{x \in \tilde{V}(\Sigma) : \langle x, y \rangle_\Sigma = 0 \; \forall \; y \in$

$\tilde{V}(\Sigma)\}$. (Strictly speaking, the set defined above should be called the 'left'-radical of the form; but for hermitian forms, the left- and right- radicals clearly agree.)

Define $V(\Sigma) = \tilde{V}(\Sigma)/\mathcal{N}(\Sigma)$, and $\zeta_M = [M] + \mathcal{N}(\Sigma)$. It should be clear that $\langle \cdot, \cdot \rangle_\Sigma$ descends to a non-degenerate hermitian sesquilinear form on $V(\Sigma)$ and that $V(\Sigma)$ is indeed generated, as a $k$-module, by $\{\zeta_M : M \in Hom(\emptyset, \Sigma)\}$.

Note that if $M \in Hom(\Sigma_1, \Sigma_2)$, then the equation

$$\widetilde{Z_M}([M_1]) = [M_1 \circ M]$$

clearly defines a (unique) $k$-homomorphism $\widetilde{Z_M} : \tilde{V}(\Sigma_1) \to \tilde{V}(\Sigma_2)$. We assert that $\widetilde{Z_M}(\mathcal{N}(\Sigma_1)) \subseteq \mathcal{N}(\Sigma_2)$. (*Reason:* If $\sum_{M_1} \alpha_{M_1}[M_1] \in \mathcal{N}(\Sigma_1)$ is arbitrary, then for any $[M_2] \in Hom(\emptyset, \Sigma_2)$, we have

$$\begin{aligned}
&\langle \widetilde{Z_M}\left(\sum_{M_1} \alpha_{M_1}[M_1]\right), [M_2]\rangle_{\Sigma_2} \\
&= \sum_{M_1} \alpha_{M_1} \langle \widetilde{Z_M}([M_1]), [M_2]\rangle\rangle_{\Sigma_2} \\
&= \sum_{M_1} \alpha_{M_1} \langle [M \circ M_1], [M_2]\rangle\rangle_{\Sigma_2} \\
&= \sum_{M_1} \alpha_{M_1} < M_2^* \circ M \circ M_1 > \\
&= \sum_{M_1} \alpha_{M_1} < ( M^* \circ M_2 )^* \circ M_1 > \\
&= \sum_{M_1} \alpha_{M_1} \langle [M_1], [M^* \circ M_2]\rangle\rangle_{\Sigma_1} \\
&= \langle (\sum_{M_1} \alpha_{M_1}[M_1]), [M^* \circ M_2]\rangle\rangle_{\Sigma_1} \\
&= 0 \,.)
\end{aligned}$$

Hence, there exists a unique (well-defined) $k$-homomorphism $Z_M : V(\Sigma_1) \to V(\Sigma_2)$ such that

$$Z_M(\zeta_{M_1}) = \zeta_{M \circ M_1} \;\forall\; M_1 \in Hom(\emptyset, \Sigma_1) \,.$$

It is easily verified now that $V : \mathcal{C} \to k - mod$ is a functor, and that equation (1.2.7) is satisfied.

We need to verify conditions (Q1) and (Q2). As for (Q1), consider the $k$-linear map $f : \tilde{V}(\emptyset) \to k$ defined by

$$f(\sum_M \alpha_M[M]) = \sum_M \alpha_M < M > ,$$

and note that

$$\begin{aligned}\sum_M \alpha_M[M] \in \mathcal{N}(\emptyset) &\Leftrightarrow \sum_M \alpha_M \langle [M],[N]\rangle_\emptyset = 0 \; \forall \; N \\ &\Leftrightarrow \sum_M \alpha_M < N^* \circ M > = 0 \; \forall \; N \\ &\Leftrightarrow \sum_M \alpha_M < N^* > \; < M > = 0 \; \forall \; N \\ &\Leftrightarrow \sum_M \alpha_M < M > = 0 \\ &\Leftrightarrow f(\sum_M \alpha_M[M]) = 0 ;\end{aligned}$$

i.e., $\mathcal{N}(\emptyset) = ker \; f$. Hence the map $[M] \mapsto f(M)$ descends to an isomorphism $V(\emptyset) \cong k$ under which $\zeta_M$ gets identified with $< M >$ for every $M \in Hom(\emptyset, \emptyset)$. It follows that $V$ satisfies (Q1) and that $<>_V = <>$.

The definitions of the form $\langle \cdot, \cdot \rangle_\Sigma$ and of $V(\Sigma)$ ensure that $V$ satisfies (Q2) also, and is hence a c.g.q.f.

If $(V', Z', \zeta')$ is another c.g.q.f. such that $<>_{V'} = <>_V$, it is a consequence of Remark 1.2.5 that there exists, for each $\Sigma \in Obj(\mathcal{C})$, a unique $k$-module isomorphism $f_\Sigma : V(\Sigma) \to V'(\Sigma)$ such that $f_\Sigma(\zeta_M) = \zeta'_M \; \forall M \in Hom(\emptyset, \Sigma)$, and hence, $Z'_M = f_{\Sigma_2} Z'_M f_{\Sigma_1}^{-1} \; \forall M \in Hom(\Sigma_1, \Sigma_2)$; thus the two functors are naturally equivalent.

□

REMARK 1.2.5 Suppose that, for $i = 1, 2$, there exists a $k$-module $V_i$ which is equipped with a non-degenerate hermitian sesquilinear form $\langle \cdot, \cdot \rangle_{(i)}$, and that $T_i = \{x_\alpha^{(i)} : \alpha \in \Lambda\}$ is a total set in $V_i$ - meaning that the submodule generated by $T_i$ is all of $V_i$. (Note that the index set $\Lambda$ is the same for both sets.) Suppose, in addition, that $\langle x_\alpha^{(1)}, x_\beta^{(1)} \rangle_{(1)} = \langle x_\alpha^{(2)}, x_\beta^{(2)} \rangle_{(2)}$ , $\forall \; \alpha, \beta \in$

$\Lambda$. Then there exists a unique $k$-module isomorphism $f : V_1 \to V_2$ such that $f(x_\alpha^{(1)}) = x_\alpha^{(2)}$ , $\forall \alpha \in \Lambda$. (*Reason:* note that

$$\begin{aligned}
\sum_{i=1}^{n} c_i x_{\alpha_i}^{(1)} = 0 \;&\Leftrightarrow\; \langle \sum_{i=1}^{n} c_i x_{\alpha_i}^{(1)}, \sum_{j=1}^{m} d_j x_{\beta_j}^{(1)} \rangle_{(1)} = 0 \;,\; \forall \sum_{j=1}^{m} d_j x_{\beta_j}^{(1)} \\
&\Leftrightarrow\; \langle \sum_{i=1}^{n} c_i x_{\alpha_i}^{(2)}, \sum_{j=1}^{m} d_j x_{\beta_j}^{(2)} \rangle_{(2)} = 0 \;,\; \forall \sum_{j=1}^{m} d_j x_{\beta_j}^{(2)} \\
&\Leftrightarrow\; \sum_{i=1}^{n} c_i x_{\alpha_i}^{(2)} = 0 \;;
\end{aligned}$$

deduce the existence of the desired isomorphism $f$.)

REMARK 1.2.6 Since the objects of $\mathcal{C}$ will be 'marked' smooth oriented $d$-folds, while the morphisms of $\mathcal{C}$ will be isomorphism classes, rel boundary, of 'marked' smooth oriented $(d+1)$-folds viewed as equivalence classes of cobordisms, we have the following additional structure on our cobordism categories. Namely, if $M$ is a smooth oriented $(d+1)$-fold with $\partial M = \Sigma_1^* \coprod \Sigma_2$, we may view $M$ as a cobordism from $\Sigma_1$ to $\Sigma_2$, or as a cobordism from $\emptyset$ to $\Sigma_1^* \coprod \Sigma_2$. Hence we have a bijection from $Hom(\Sigma_1, \Sigma_2)$ to $Hom(\emptyset, \Sigma_1^* \coprod \Sigma_2)$, which we shall denote by

$$M_{\Sigma_1,\Sigma_2} \mapsto M_{\emptyset,\Sigma_1^* \coprod \Sigma_2} \;.$$

We shall use the above notation when necessary to avoid possible confusion.

If $M, N \in Hom(\emptyset, \Sigma)$, then the following morphisms make sense:

$$M_{\Sigma^*,\emptyset},\; M^*_{\Sigma,\emptyset},\; M^*_{\emptyset,\Sigma^*} ,N_{\Sigma^*,\emptyset},\; N^*_{\Sigma,\emptyset},\; N^*_{\emptyset,\Sigma^*} \;;$$

and we have:

$$N^*_{\Sigma,\emptyset} \circ M_{\emptyset,\Sigma} \;=\; (N_{\Sigma^*,\emptyset} \circ M^*_{\emptyset,\Sigma^*})^* \;. \qquad (1.2.9)$$

(This is because, as cobordisms, the left- and right- sides correspond to the smooth oriented $(d+1)$-folds $M \cup_\Sigma N$ and $(M^* \cup_{\Sigma^*} N^*)^*$, respectively.)

PROPOSITION 1.2.7 *Suppose* $V : \mathcal{C} \to k-mod$ *is a c.g.q.f. Let* $\Sigma, \Sigma_1, \Sigma_2 \in Obj(\mathcal{C})$.

*(1) There exists a unique k-linear monomorphism*

$$D_\Sigma : V(\Sigma^*) \to V(\Sigma)^* \ (= Hom(V(\Sigma), k))$$

*such that*

$$\Big( D_\Sigma(\zeta_{M^*_{\emptyset,\Sigma^*}}) \Big)(\zeta_N) \ = \ \langle \zeta_N, \zeta_M \rangle_\Sigma \ \forall M, N \in Hom(\emptyset, \Sigma) \ . \tag{1.2.10}$$

*(2) There exists a unique k-module homomorphism*

$$\mu_{\Sigma_1,\Sigma_2} : V(\Sigma_1) \otimes_k V(\Sigma_2) \to V(\Sigma_1 \coprod \Sigma_2)$$

*such that*

$$\mu_{\Sigma_1,\Sigma_2}(\zeta_{M_1} \otimes \zeta_{M_2}) \ = \ \zeta_{M_1 \coprod M_2} \ \forall M_i \in Hom(\emptyset, \Sigma_i) \ .$$

**Proof:** (1) Since $\{[M^*_{\emptyset,\Sigma^*}] : M \in Hom(\emptyset, \Sigma)\}$ is clearly a basis for $\tilde{V}(\Sigma^*)$, there is a unique $k$-homomorphism $D : \tilde{V}(\Sigma^*) \to \tilde{V}(\Sigma)^*$ such that $(D([M^*_{\emptyset,\Sigma^*}]))([N_{\emptyset,\Sigma}]) = \langle \zeta_N, \zeta_M \rangle_\Sigma \ \forall M, N \in Hom(\emptyset, \Sigma)$ . Clearly $D([M^*_{\emptyset,\Sigma^*}])(\mathcal{N}(\Sigma)) = 0 \ \forall M$, so that $D : \tilde{V}(\Sigma^*) \to V(\Sigma)^*$. Note that

$$\begin{aligned}
&\sum_M \alpha_M [M^*] \in \mathcal{N}(\Sigma^*) \\
\Leftrightarrow \ & \sum_M \alpha_M \langle [M^*], [N^*] \rangle_{\Sigma^*} = 0 \ , \ \forall \ [N_{\emptyset,\Sigma}] \\
\Leftrightarrow \ & \sum_M \alpha_M < N_{\Sigma^*,\emptyset} \circ M^*_{\emptyset,\Sigma^*} >_V = 0 \ , \ \forall \ [N_{\emptyset,\Sigma}] \\
\Leftrightarrow \ & \sum_M \alpha_M \overline{< N^*_{\Sigma,\emptyset} \circ M_{\emptyset,\Sigma} >_V} = 0 \ , \ \forall \ [N_{\emptyset,\Sigma}] \quad \text{(by eqn. (1.2.9))} \\
\Leftrightarrow \ & \sum_M \alpha_M \overline{\langle [M_{\emptyset,\Sigma}], [N_{\emptyset,\Sigma}] \rangle_\Sigma} = 0 \ , \ \forall \ [N_{\emptyset,\Sigma}] \\
\Leftrightarrow \ & \sum_M \alpha_M \langle [N_{\emptyset,\Sigma}], [M_{\emptyset,\Sigma}] \rangle_\Sigma = 0 \ , \ \forall \ [N_{\emptyset,\Sigma}] \\
\Leftrightarrow \ & (D(\sum_M \alpha_M [M^*]))([N]) = 0 \ , \ \forall \ [N_{\emptyset,\Sigma}] \\
\Leftrightarrow \ & D(\sum_M \alpha_M [M^*]) = 0 \ .
\end{aligned}$$

Conclude that $ker\ D = \mathcal{N}(\Sigma^*)$ and hence that $D$ descends to a $k$-monomorphism $D_\Sigma : V(\Sigma^*) \to V(\Sigma)^*$ which clearly maps $\zeta_{M^*_{\emptyset,\Sigma^*}} = [M^*_{\emptyset,\Sigma^*}] + \mathcal{N}(\Sigma^*)$ to $\langle \cdot, \zeta_M \rangle_\Sigma$.

(2) Clearly there exists a $k$-linear map $\tilde{\mu} : \tilde{V}(\Sigma_1) \otimes \tilde{V}(\Sigma_2) \to V(\Sigma_1 \coprod \Sigma_2)$ defined by

$$\tilde{\mu}([M_1] \otimes [M_2]) = \zeta_{M_1 \coprod M_2} .$$

We need to show that this $\tilde{\mu}$ 'factors through' $(V(\Sigma_1) \otimes V(\Sigma_2))$. For this, we should verify that $ker(\tilde{\mu}) \supseteq ((\mathcal{N}(\Sigma_1) \otimes \tilde{V}(\Sigma_2)) \cup (\tilde{V}(\Sigma_1) \otimes \mathcal{N}(\Sigma_2)))$. We shall check that $ker(\tilde{\mu}) \supseteq \mathcal{N}(\Sigma_1) \otimes \tilde{V}(\Sigma_2)$, the other verification being similar.

Let $\sum_i \alpha_{M_1^{(i)}} [M_1^{(i)}] \in \mathcal{N}(\Sigma_1)$ (where $M_1^{(i)} \in Hom(\emptyset, \Sigma_1)\ \forall i$), and let $M_2 \in Hom(\emptyset, \Sigma_2)$ be arbitrary.

Since the form $\langle \cdot, \cdot \rangle_{\Sigma_1 \coprod \Sigma_2}$ is non-degenerate, it suffices to verify that $\langle \tilde{\mu}((\sum_i \alpha_{M_1^{(i)}} [M_1^{(i)}]) \otimes [M_2]), \zeta_M \rangle_{\Sigma_1 \coprod \Sigma_2} = 0$ for every $M \in Hom(\emptyset, \Sigma_1 \coprod \Sigma_2)$. This is true because

$$\begin{aligned}
&\langle \tilde{\mu}((\sum_i \alpha_{M_1^{(i)}} [M_1^{(i)}]) \otimes [M_2]), \zeta_M \rangle_{\Sigma_1 \coprod \Sigma_2} \\
&= \sum_i \alpha_{M_1^{(i)}} \langle \zeta_{M_1^{(i)} \coprod M_2}, \zeta_M \rangle_{\Sigma_1 \coprod \Sigma_2} \\
&= \sum_i \alpha_{M_1^{(i)}} < (M_1^{(i)} \coprod M_2) \cup_{\Sigma_1 \coprod \Sigma_2} M^* >_V \\
&= \sum_i \alpha_{M_1^{(i)}} < M_1^{(i)} \cup_{\Sigma_1} (M \cup_{\Sigma_2} M_2^*)^* >_V \\
&= \langle (\sum_i \alpha_{M_1^{(i)}} [M_1^{(i)}]), [M \cup_{\Sigma_2} M_2^*] \rangle_{\Sigma_1} \\
&= 0 \quad (\text{by definition of } \mathcal{N}(\Sigma_1)\ ) ,
\end{aligned}$$

and the proof of the proposition is complete. □

Consider the following requirements on a c.g.q.f. $(V, Z, \zeta)$.

**(I)** The map $D_\Sigma$ of Proposition 1.2.7(1) is an isomorphism of $k$-modules, for each $\Sigma$. (Thus, it is required to also be surjective.)

**(M)** The map $\mu_{\Sigma_1,\Sigma_2}$ of Proposition 1.2.7(2) is an isomorphism of $k$-modules, for all $\Sigma_1, \Sigma_2$.

**(F)** For each $\Sigma$, the module $V(\Sigma)$ is a free module of finite rank, and further, the form $\langle \cdot, \cdot \rangle_\Sigma$ is *unimodular* - meaning that the matrix $((\langle e_i, e_j \rangle_\Sigma))$ is invertible for some (equivalently every) basis $\{e_i\}$ of $V(\Sigma)$.

It is easy to verify that condition (F) implies condition (I). A **topological quantum field theory** (TQFT in the sequel) is a quantisation functor which satisfies the above requirements or some variants thereof. Before getting to the general definition, we shall first define what it means for what we call a c.g.q.f to be a TQFT.

DEFINITION 1.2.8 *A (d+1)-dimensional c.g. TQFT is a c.g.q.f. which satisfies conditions (M) and (F) above.*

PROPOSITION 1.2.9 *Suppose V is a c.g. TQFT.*
*If* $M, N \in Hom(\Sigma_1, \Sigma_2)$, *then*

$$\begin{aligned} \langle \zeta_{M_{\emptyset,\Sigma_1^* \coprod \Sigma_2}}, \zeta_{N_{\emptyset,\Sigma_1^* \coprod \Sigma_2}} \rangle_{\Sigma_1^* \coprod \Sigma_2} &= trace(Z^*_{N_{\Sigma_1,\Sigma_2}} Z_{M_{\Sigma_1,\Sigma_2}}) \\ &= trace(Z_{M_{\Sigma_1,\Sigma_2}} Z^*_{N_{\Sigma_1,\Sigma_2}}) \ , \end{aligned}$$

*where* $T^*$ *denotes the 'adjoint' with respect to the forms* $\langle \cdot, \cdot \rangle_{\Sigma_i}$.

**Proof:** For $M = M_{\emptyset,\Sigma_1^* \coprod \Sigma_2} \in Hom(\emptyset, \Sigma_1^* \coprod \Sigma_2)$, we assert that

$$\langle \zeta_{M_{\emptyset,\Sigma_1^* \coprod \Sigma_2}}, \zeta_{M_1^* \coprod M_2} \rangle_{\Sigma_1^* \coprod \Sigma_2} = \langle Z_{M_{\Sigma_1,\Sigma_2}}(\zeta_{M_1}), \zeta_{M_2} \rangle_{\Sigma_2} \ , \quad (1.2.11)$$

for all $M_i \in Hom(\emptyset, \Sigma_i)$, $i = 1, 2$. (This is because

$$\begin{aligned} \langle \zeta_{M_{\emptyset,\Sigma_1^* \coprod \Sigma_2}}, \zeta_{M_1^* \coprod M_2} \rangle_{\Sigma_1^* \coprod \Sigma_2} &= <(M \cup_{\Sigma_1^* \coprod \Sigma_2} (M_1 \coprod M_2^*) >_V \\ &= <(M \cup_{\Sigma_1^*} M_1) \cup_{\Sigma_2} M_2^* >_V \\ &= \langle \zeta_{M \cup_{\Sigma_1^*} M_1}, \zeta_{M_2} \rangle_{\Sigma_2} \\ &= \langle \zeta_{M_{\Sigma_1,\Sigma_2} \circ M_1}, \zeta_{M_2} \rangle_{\Sigma_2} \\ &= \langle Z_{M_{\Sigma_1,\Sigma_2}}(\zeta_{M_1}), \zeta_{M_2} \rangle_{\Sigma_2} \ . \ ) \end{aligned}$$

The non-degeneracy of the form $\langle \cdot, \cdot \rangle_{\Sigma_2}$ implies, via equation (1.2.11), that

$$\sum_M \alpha_M \zeta_M = 0 \ \Rightarrow \ \sum_M \alpha_M Z_M = 0.$$

So we have a map $\Psi : V(\Sigma_1^* \coprod \Sigma_2) \to \mathrm{Hom}(V(\Sigma_1), V(\Sigma_2))$ such that

$$\Psi(\zeta_{M_{\emptyset,\Sigma_1^* \coprod \Sigma_2}}) = Z_{M_{\Sigma_1,\Sigma_2}} \quad \forall\, M \in Hom(\Sigma_1, \Sigma_2).$$

Setting $M = N_1^* \coprod N_2$, we see from equation (1.2.11) that

$$\begin{aligned}
&\langle (\Psi(\zeta_{N_1^* \coprod N_2}))(\zeta_{M_1}), \zeta_{M_2} \rangle_{\Sigma_2} \\
&= \langle \zeta_{N_1^* \coprod N_2}, \zeta_{M_1^* \coprod M_2} \rangle_{\Sigma_1^* \coprod \Sigma_2} \\
&= < (N_1^* \coprod N_2) \cup_{\Sigma_1^* \coprod \Sigma_2} (M_1 \coprod M_2^*) >_V \\
&= < (N_1^* \cup_{\Sigma_1^*} M_1) \coprod (N_2 \cup_{\Sigma_2} M_2^*) >_V \\
&= < (N_1^* \cup_{\Sigma_1^*} M_1) >_V < (N_2 \cup_{\Sigma_2} M_2^*) >_V \\
&= \langle \zeta_{M_1}, \zeta_{N_1} \rangle_{\Sigma_1} \langle \zeta_{N_2}, \zeta_{M_2} \rangle_{\Sigma_1} ,
\end{aligned}$$

and hence

$$\Psi(\zeta_{N_1^* \coprod N_2}) = \langle \cdot, \zeta_{N_1} \rangle_{\Sigma_1} \zeta_{N_2} \, . \tag{1.2.12}$$

Next, consider the following chain of isomorphisms:

$$\begin{aligned}
V(\Sigma_1^* \coprod \Sigma_2) &\overset{\mu^{-1}_{\Sigma_1^*,\Sigma_2}}{\cong} V(\Sigma_1^*) \otimes V(\Sigma_2) \\
&\overset{D_{\Sigma_1} \otimes id_{V(\Sigma_2)}}{\cong} V(\Sigma_1)^* \otimes V(\Sigma_2) \\
&\overset{T}{\cong} \mathrm{Hom}_k(V(\Sigma_1), V(\Sigma_2)) ,
\end{aligned}$$

where $(T(f \otimes w))(v) = f(v)w$. Under these isomorphisms, it is seen that the following vectors correspond to one another, for any $M_i \in Hom(\emptyset, \Sigma_i)$:

$$\zeta_{M_1^* \coprod M_2} \leftrightarrow \zeta_{M_1^*} \otimes \zeta_{M_2} \leftrightarrow \langle \cdot, \zeta_{M_1} \rangle_{\Sigma_1} \otimes \zeta_{M_2} \leftrightarrow \langle \cdot, \zeta_{M_1} \rangle_{\Sigma_1} \zeta_{M_2} \, .$$

So the equation

$$\Phi(\zeta_{M_1^* \coprod M_2}) = \langle \cdot, \zeta_{M_1} \rangle_{\Sigma_1} \zeta_{M_2} \tag{1.2.13}$$

defines an isomorphism $\Phi : V(\Sigma_1^* \coprod \Sigma_2) \to \mathrm{Hom}(V(\Sigma_1), V(\Sigma_2))$.

Since vectors such as $\zeta_{M_1^* \coprod M_2}$ form a total set in $V(\Sigma_1^* \coprod \Sigma_2)$, we see, on comparing equations (1.2.12) and (1.2.13), that $\Phi = \Psi$. We conclude, in particular, that $\Phi(\zeta_{M_{\emptyset,\Sigma_1^* \coprod \Sigma_2}}) = Z_{M_{\Sigma_1,\Sigma_2}}$.

Hence the proposition will be proved once we show that

$$\langle \xi, \eta \rangle_{\Sigma_1^* \coprod \Sigma_2} = \text{trace}(\Phi(\xi)\Phi(\eta)^*) = \text{trace}(\Phi(\eta)^*\Phi(\xi)) \; \forall \xi, \eta \, .$$

Since this latter equation depends linearly on $\xi$ and conjugate-linearly on $\eta$, it suffices to prove this for $\xi, \eta$ coming from the total set of vectors of the form $\zeta_{M_1^* \coprod M_2}$.

Setting $\xi = \zeta_{M_1^* \coprod M_2}$, $\eta = \zeta_{N_1^* \coprod N_2}$, we find that:

$$\begin{aligned}
\langle \zeta_{M_1^* \coprod M_2}, \zeta_{N_1^* \coprod N_2} \rangle_{\Sigma_1^* \coprod \Sigma_2} &= \langle \zeta_{N_1}, \zeta_{M_1} \rangle_{\Sigma_1} \langle \zeta_{M_2}, \zeta_{N_2} \rangle_{\Sigma_2} \\
\Phi(\zeta_{M_1^* \coprod M_2})\Phi(\zeta_{N_1^* \coprod N_2})^* &= (\langle \cdot, \zeta_{M_1} \rangle_{\Sigma_1} \zeta_{M_2})(\langle \cdot, \zeta_{N_1} \rangle_{\Sigma_1} \zeta_{N_2})^* \\
&= (\langle \cdot, \zeta_{M_1} \rangle_{\Sigma_1} \zeta_{M_2})(\langle \cdot, \zeta_{N_2} \rangle_{\Sigma_2} \zeta_{N_1}) \\
&= \langle \zeta_{N_1}, \zeta_{M_1} \rangle_{\Sigma_1} (\langle \cdot, \zeta_{N_2} \rangle_{\Sigma_2} \zeta_{M_2}) \\
\Phi(\zeta_{N_1^* \coprod N_2})^* \Phi(\zeta_{M_1^* \coprod M_2}) &= (\langle \cdot, \zeta_{N_1} \rangle_{\Sigma_1} \zeta_{N_2})^* (\langle \cdot, \zeta_{M_1} \rangle_{\Sigma_1} \zeta_{M_2}) \\
&= (\langle \cdot, \zeta_{N_2} \rangle_{\Sigma_2} \zeta_{N_1})(\langle \cdot, \zeta_{M_1} \rangle_{\Sigma_1} \zeta_{M_2}) \\
&= \langle \zeta_{M_2}, \zeta_{N_2} \rangle_{\Sigma_2} (\langle \cdot, \zeta_{M_1} \rangle_{\Sigma_1} \zeta_{N_1}).
\end{aligned}$$

Hence, in order to complete the proof of the theorem, it suffices to prove that, for arbitrary $\xi, \eta \in V(\Sigma)$, we have:

$$\text{trace}(\langle \cdot, \eta \rangle_\Sigma \xi) = \langle \xi, \eta \rangle_\Sigma \, .$$

Since both sides of this equation depend linearly on $\xi$, it suffices to verify this for $\xi$ belonging to a basis, say $\{\xi_1, \cdots, \xi_n\}$. So suppose $\xi = \xi_j$. Writing $\{\xi_i^*\}_{i=1}^n$ for the dual basis, we find that

$$\begin{aligned}
\text{trace}(\langle \cdot, \eta \rangle_\Sigma \xi_j) &= \sum_{i=1}^n \langle \xi_i, \eta \rangle_\Sigma \xi_i^*(\xi_j) \\
&= \langle \xi_j, \eta \rangle_\Sigma \, ,
\end{aligned}$$

as desired, and the proof is complete. □

Note the following immediate consequences of Proposition 1.2.9.

COROLLARY 1.2.10 *(a) Suppose $M \in Hom(\Sigma, \Sigma)$. Then,*

$$< M \circ id_\Sigma^* >_V = \langle \zeta_M, \zeta_{id_\Sigma} \rangle_{\Sigma^* \coprod \Sigma} = trace(Z_M) \, .$$

*(b) In particular, when* $C = C_0$, *we find that if* $M$ *is a* $(d+1)$-*fold with* $\partial M = \Sigma^* \coprod \Sigma$, *and if* $M_\Sigma$ *denotes the closed manifold obtained by identifying the two copies of* $\Sigma$ *in* $M$, *then* $< M_\Sigma >_V = trace(Z_{M_{\Sigma,\Sigma}})$.

*Again, when* $C = C_0$, $< \Sigma \times S^1 >_V$ *is the rank of the module* $V(\Sigma)$.

DEFINITION 1.2.11 *A* $(d+1)$-*dimensional* **topological quantum field theory** *(TQFT in the sequel) is a quantisation functor which satisfies the following finiteness and multiplicativity requirements:*

**(F)** *(finiteness) For each* $\Sigma$, *the module* $V(\Sigma)$ *is a free module of finite rank, and further, the form* $\langle \cdot, \cdot \rangle_\Sigma$ *is* unimodular - *meaning that the matrix* $((\langle e_i, e_j \rangle_\Sigma))$ *is invertible for some (equivalently every) basis* $\{e_i\}$ *of* $V(\Sigma)$.

**(M)** *(multiplicativity) For each* $\Sigma_1, \Sigma_2$, *there exists an isomorphism*

$$\mu_{\Sigma_1,\Sigma_2} : V(\Sigma_1) \otimes V(\Sigma_2) \rightarrow V(\Sigma_1 \coprod \Sigma_2)$$

*which is 'natural' in that it satisfies the following four requirements:*

*(i) (associativity) The following diagram of k-modules and homomorphisms is commutative, for every* $\Sigma_1, \Sigma_2, \Sigma_3$ :

$$\begin{array}{ccc} V(\Sigma_1) \otimes V(\Sigma_2) \otimes V(\Sigma_3) & \stackrel{\mu_{\Sigma_1,\Sigma_2} \otimes id_{V(\Sigma_3)}}{\rightarrow} & V(\Sigma_1 \coprod \Sigma_2) \otimes V(\Sigma_3) \\ id_{V(\Sigma_1)} \otimes \mu_{\Sigma_2,\Sigma_3} \downarrow & & \downarrow \mu_{\Sigma_1 \coprod \Sigma_2, \Sigma_3} \\ V(\Sigma_1) \otimes V(\Sigma_2 \coprod \Sigma_3) & \stackrel{\mu_{\Sigma_1,\Sigma_2 \coprod \Sigma_3)}}{\rightarrow} & V(\Sigma_1 \coprod \Sigma_2 \coprod \Sigma_3) \end{array}$$

*(In the sequel, we shall omit the subscripts for the* $\mu$-*maps if the context makes it clear what they should be.)*

*(ii) (empty set) For every* $\Sigma$, *the following composite map is equal to* $id_{V(\Sigma)}$:

$$V(\Sigma) = k \otimes V(\Sigma) = V(\emptyset) \otimes V(\Sigma) \stackrel{\mu}{\rightarrow} V(\emptyset \coprod \Sigma) = V(\Sigma)$$

*(iii) (commutativity) For every* $\Sigma_1, \Sigma_2$, *the following diagram is commutative:*

$$\begin{array}{ccc} V(\Sigma_1) \otimes V(\Sigma_2) & \stackrel{\mu}{\rightarrow} & V(\Sigma_1 \coprod \Sigma_2) \\ flip \downarrow & & \downarrow id \\ V(\Sigma_2) \otimes V(\Sigma_1) & \stackrel{\mu}{\rightarrow} & V(\Sigma_2 \coprod \Sigma_1) \end{array}$$

*(iv) (disjoint sums of bordisms) If $M_i \in Hom(\Sigma_i, \Sigma_i')$, $i = 1, 2$, the following diagram is commutative:*

$$\begin{array}{ccc} V(\Sigma_1) \otimes V(\Sigma_2) & \stackrel{\mu}{\to} & V(\Sigma_1 \coprod \Sigma_2) \\ Z_{M_1} \otimes Z_{M_2} \downarrow & & \downarrow Z_{M_1 \coprod M_2} \\ V(\Sigma_1') \otimes V(\Sigma_2') & \stackrel{\mu}{\to} & V(\Sigma_1' \coprod \Sigma_2') \end{array}$$

□

## 1.3 Fusion algebras

DEFINITION 1.3.1 *A* **fusion algebra** *is a finite-dimensional associative, unital, involutive algebra $\mathcal{A}$ over $\mathbb{C}$ equipped with a distinguished basis $\mathcal{B} = \mathcal{B}_{\mathcal{A}}$, such that the structure constants $\{N_{XY}^Z : X, Y, Z \in \mathcal{B}\}$ defined by*

$$XY = \sum_{Z \in \mathcal{B}} N_{XY}^Z \, Z \tag{1.3.14}$$

*satisfy the following conditions:*

*(i)* $N_{XY}^Z \in \mathbb{Z}_+ \ (= \{0, 1, 2, \cdots\})$ , $\forall \ X, Y, Z$ ;

*(ii) the identity, denoted simply by 1, of the algebra $\mathcal{A}$, belongs to $\mathcal{B}$; then, clearly*

$$N_{X1}^Y = N_{1X}^Y = \delta_X^Y \ , \ \forall \ X, Y \in \mathcal{B} \ ;$$

*(iii) $\mathcal{B}_{\mathcal{A}}$ is closed under the involution of $\mathcal{A}$; hence, there exists an involution $\mathcal{B} \ni X \mapsto \bar{X} \in \mathcal{B}$ such that $\bar{X} = X^* \ \forall X \in \mathcal{B}$.*

*(iv)* $N_{XY}^Z = N_{\bar{X}Z}^Y \ \forall \ X, Y, Z \in \mathcal{B}$.

We first list some easy consequences of the axioms.

REMARK 1.3.2 (a)

$$N_{XY}^1 = N_{\bar{X}1}^Y = \delta_{\bar{X},Y} \ , \ \forall X, Y \in \mathcal{B} \ . \tag{1.3.15}$$

(b) We shall write $\mathcal{A} \ni x \mapsto L_x \in End_{\mathbb{C}}(\mathcal{A})$ to denote the left regular representation (defined by $L_x(y) = xy$). The map $L$ is one-to-one since $\mathcal{A}$ has an identity.

Equip $\mathcal{A}$ with an inner product such that $\mathcal{B}_{\mathcal{A}}$ is an orthonormal basis. We then identify ($\mathbb{C}$-linear) endomorphisms of $\mathcal{A}$ and their matrices with respect to this basis. Thus, $End_{\mathbb{C}}(\mathcal{A}) \ni T \leftrightarrow ((t_{XY})) \in M_{\mathcal{B}}(\mathbb{C})$, where $t_{XY} = \langle TY, X\rangle \ \forall X, Y \in \mathcal{B}$.

(c) With the conventions of (b) above, we assert that $x \mapsto L_x$ is a *-homomorphism (in fact *-monomorphism, by (b) above) of $\mathcal{A}$ into $M_{\mathcal{B}}(\mathbb{C})$, and consequently $\mathcal{A}$ has the structure of a $C^*$-algebra.

(*Reason:* It suffices to verify that $L_{\bar{X}} = (L_X)^* \ \forall X$. This is true because

$$
\begin{aligned}
(L_{\bar{X}})_{YZ} &= \langle \bar{X}Z, Y\rangle \\
&= N^{Y}_{\bar{X}Z} \\
&= N^{Z}_{XY} && \text{(by Definition 1.3.1(iv))} \\
&= \overline{N^{Z}_{XY}} && \text{(since } t = \bar{t} \text{ for real } t) \\
&= \overline{\langle XY, Z\rangle} \\
&= \overline{(L_X)_{ZY}} \\
&= (L_X)^*_{YZ} \ .)
\end{aligned}
$$

(d) Consider the map $\tau : \mathcal{A} \to \mathbb{C}$ defined by

$$
\tau(x) \;=\; \langle x, 1\rangle \;=\; \langle L_x 1, 1\rangle \ .
$$

This is clearly a linear functional. Further, if $x = \sum_X a_X X$, $y = \sum_Y b_Y Y$, then $y^* = \sum_Y \overline{b_Y}\bar{Y} = \sum_X \overline{b_{\bar{X}}} X$, and hence

$$
\begin{aligned}
\tau(yx) &= \langle L_{yx} 1, 1\rangle \\
&= \langle L_y L_x 1, 1\rangle \\
&= \langle L_x 1, (L_y)^* 1\rangle \\
&= \langle L_x 1, L_{y^*} 1\rangle \\
&= \langle x, y^*\rangle \\
&= \sum_X a_X b_{\bar{X}} \\
&= \sum_Z b_Z a_{\bar{Z}} \\
&= \tau(xy) \ ,
\end{aligned}
$$

and hence $\tau$ is a *tracial* linear functional.

Note that the above computation also shows that

$$\tau(y^*x) = \langle x, y \rangle \ . \tag{1.3.16}$$

In particular, $\tau(x^*x) = ||x||^2$; hence, $\tau$ is a *positive* functional on the $C^*$-algebra $\mathcal{A}$ (which just means that it assumes non-negative values on the 'positive elements' of $\mathcal{A}$, i.e., those elements of the form $x^*x$), which is *faithful* in the sense that $x \neq 0 \Rightarrow \tau(x^*x) > 0$.

Thus, $\tau$ is a faithful, positive, tracial linear functional on $\mathcal{A}$, 'normalised' so that it attains the value 1 at the identity of $\mathcal{A}$ (since $||X_0|| = 1$).

(e) We also have the following 'right-sided' analogue of condition (iv) in the definition of a fusion algebra:

$$N_{XY}^{Z} \ = \ N_{Z\bar{Y}}^{X} \ . \tag{1.3.17}$$

This is because

$$\langle XY, Z \rangle = \tau(Z^*XY) = \tau(YZ^*X) = \tau((Z\bar{Y})^*X) = \langle X, Z\bar{Y} \rangle \ .$$

(f) Combining (e) with condition (iv) in the definition of a fusion algebra, we see thus that, for all $X, Y, Z \in \mathcal{B}$,

$$N_{XY}^{Z} \ = \ N_{\bar{X}Z}^{Y} \ = \ N_{Z\bar{Y}}^{X} \ = \ N_{Y\bar{Z}}^{\bar{X}} \ = \ N_{\bar{Z}X}^{\bar{Y}} \ = \ N_{\bar{Y}\bar{X}}^{\bar{Z}} \ .$$

We come next to the important notion of *dimension function* on a fusion algebra.

PROPOSITION 1.3.3 *There exists a unique function $d : \mathcal{A} \to \mathbb{C}$ which satisfies the following conditions:*

*(i) d is an algebra homomorphism; equivalently,*

$$d(X)d(Y) \ = \ \sum_{Z} N_{XY}^{Z} d(Z) \ , \ \forall X, Y \in \mathcal{B} \ ; \ and \tag{1.3.18}$$

*(ii)* $d(X) > 0 \ \forall X \in \mathcal{B}$.

*This function $\mathcal{B}_A \ni X \mapsto d(X)$ is called* **the dimension function** *of the fusion algebra.*

**Proof:** *Existence:* Let $x \mapsto R_x$ denote the 'right-regular' representation of $\mathcal{A}$; i.e., $R_x(y) = yx$. Let $R = \sum_Z R_Z$; thus, the matrix entries of $R$ are given by

$$\begin{aligned} R_{XY} &= \sum_Z \langle R_Z(Y), X\rangle = \sum_Z \langle YZ, X\rangle \\ &= \sum_Z N_{YZ}^X = \sum_Z N_{\bar{Y}X}^Z . \end{aligned}$$

We wish to observe next that $R_{XY} > 0 \; \forall X, Y \in \mathcal{B}$. In view of the above equation, and the non-negativity of the structure constants, we need to show that for any fixed $X, Y$, it cannot be the case that $N_{\bar{Y}X}^Z = 0 \; \forall Z$. Suppose, if possible, that for some $X, Y$, we have $N_{\bar{Y}X}^Z = 0 \; \forall Z$; this means that $\bar{Y}X = 0$, which would imply that

$$\begin{aligned} 0 &= \langle Y(\bar{Y}X), X\rangle = \langle (Y\bar{Y})X, X\rangle \\ &= \sum_Z N_{Y\bar{Y}}^Z N_{ZX}^X \geq N_{Y\bar{Y}}^1 N_{1X}^X = 1 ; \end{aligned}$$

the contradiction shows that indeed $R_{XY} > 0 \; \forall X, Y$.

Hence $R$ is a strictly positive matrix. It follows from the Perron-Frobenius theorem that $R$ has a postive eigenvalue $\lambda$ with 1-dimensional eigenspace spanned by a vector $v$ with strictly positive entries. The vector $v$ is clearly unique if we normalise it so that its 1-th co-ordinate is 1. Thus, we may assume that $v = \sum_X d_X X$, with $d_1 = 1, d_X > 0 \; \forall X$.

On the other hand, since left multiplications commute with right multiplications, we see that each $L_Y$ commutes with $R$ and consequently leaves $\mathbb{C}v = ker(R - \lambda)$ invariant; thus there exists $\lambda_Y \in \mathbb{C}$ such that $L_Y v = \lambda_Y v$ for each $Y$. Since $L_Y$ and $v$ have non-negative entries, it is clear that $\lambda_Y \geq 0$. On the other hand, our proof of the positivity of $R$ shows that $\bar{Y}X \neq 0$ for each $X$; i.e., the $X$-th column of $L_{\bar{Y}} (= (L_Y)^*)$ is not identically zero, for each $X$; hence each row of $L_Y$ is non-zero (and non-negative). Since $v$ has strictly positive entries, it follows that each entry of $L_Y v$ is strictly positive; hence $\lambda_Y > 0 \; \forall Y$.

Finally, since $\mathbb{C}v$ is invariant under $L_Y$ for each $Y \in \mathcal{B}$, it follows that $\mathbb{C}v$ is invariant under $\{L_x : x \in \mathcal{A}\}$ and that there exists a map $d : \mathcal{A} \to \mathbb{C}$ such that $L_x v = d(x)v$ for all

$x \in \mathcal{A}$. It is clear that $d$ is an algebra homomorphism, and that $d(Y) = \lambda_Y > 0\ \forall Y$, and the proof of existence is complete.

*Uniqueness:* Suppose $X \mapsto \tilde{d}(X)$ is another dimension function for $\mathcal{A}$. Since $X_0$ is the identity of $\mathcal{A}$, it is clear that $\tilde{d}(1) = 1$. Let $\tilde{v} = \sum_X \tilde{d}(X)\ X$, and note that

$$\begin{aligned} R\tilde{v} &= \sum_X \tilde{d}(X)R(X) \\ &= \sum_{X,Y} \tilde{d}(X)XY \\ &= \sum_{X,Y,Z} \tilde{d}(X)N_{XY}^{Z} Z \\ &= \sum_Z \left( \sum_{X,Y} N_{Z\bar{Y}}^{X} \tilde{d}(X) \right) Z \\ &= \sum_Z \left( \sum_Y \tilde{d}(Z)\tilde{d}(\bar{Y}) \right) Z \\ &= (\sum_Y \tilde{d}(\bar{Y}))\tilde{v}\ . \end{aligned}$$

We claim now that $(\sum_Y \tilde{d}(\bar{Y}))$ is *the* Perron-Frobenius eigenvalue of $R$. (*Reason:* First note that $R = R^*$, since it follows from equation 1.3.17 that $(R_X)^* = R_{\bar{X}}$. Since eigenvectors corresponding to distinct eigenvalues of a self-adjoint matrix are orthogonal, and since two vectors with strictly positive coordinates cannot be orthogonal, our claim follows.)

Next, since the Perron-Frobenius eigenvector of a strictly positive matrix is unique up to scaling by a positive constant, and since $d(1) = \tilde{d}(1)$, we conclude that $v = \tilde{v}$; i.e., $d(X) = \tilde{d}(X)$ for all $X$.

Hence we see that any dimension function gives precisely the coordinates of the Perron-Frobenius eigenvector of $R$, so normalised to have 1 in the 1-th coordinate. In particular, we see from the existence part that we should have $\lambda_X = d(X) = \tilde{d}(X)$, and the proof is complete.

□

REMARK 1.3.4 (a) If we write $\phi : \mathcal{A} \to \mathbb{C}$ for the dimension function, then it is easily verified (since $\mathbb{C}$ is commutative, even if $\mathcal{A}$ might not be) that the function $\phi^{\#} : \mathcal{A} \to \mathbb{C}$ defined by $\phi^{\#}(x) = \overline{\phi(x^*)}$ also defines a dimension function. Hence, by uniqueness, we find that $\phi^{\#} = \phi$. Thus,

$$d(\bar{X}) = d(X) \ \forall X \in \mathcal{B}_{\mathcal{A}} \ . \tag{1.3.19}$$

(b) The **weight** of the fusion algebra $\mathcal{A}$ is defined by

$$w \ (= \ w(\mathcal{A}) \ ) \ = \sum_X d(X)^2 \ . \tag{1.3.20}$$

For future reference, we wish to note the following identity involving the weight of a fusion algebra, valid for all $Z \in \mathcal{B}_{\mathcal{A}}$:

$$\begin{aligned} \sum_{X,Y} N_{XY}^{Z} d(X) d(Y) &= \sum_X \left( \sum_Y N_{\bar{X}Z}^{Y} d(Y) \right) d(X) \\ &= \sum_X \left( d(\bar{X}) d(Z) \right) d(X) \\ &= w d(Z) \ . \end{aligned} \tag{1.3.21}$$

(c) Note the following pleasant feature thrown up by the proof: All the $L_X$'s share a common Perron-Frobenius eigenvector, and if this vector is normalised to have 1 in the 1-th coordinate, then the $X$-th coordinate of the vector is also the Perron-Frobenius eigenvalue of $L_X$.

All this is nicely illustrated by an example, which might be called 'the $A_4$ fusion algebra'. This is a 4-dimensional fusion algebra, with $\mathcal{B}_{\mathcal{A}} = \{1, \alpha, \beta, \epsilon\}$, and the matrices corresponding to left-multiplication by the basis vectors are given by

$$L_1 = \begin{bmatrix} 1 & 0 & 0 & 0 \\ 0 & 1 & 0 & 0 \\ 0 & 0 & 1 & 0 \\ 0 & 0 & 0 & 1 \end{bmatrix}, L_\alpha = \begin{bmatrix} 0 & 1 & 0 & 0 \\ 1 & 0 & 1 & 0 \\ 0 & 1 & 0 & 1 \\ 0 & 0 & 1 & 0 \end{bmatrix},$$

$$L_\beta = \begin{bmatrix} 0 & 0 & 1 & 0 \\ 0 & 1 & 0 & 1 \\ 1 & 0 & 1 & 0 \\ 0 & 1 & 0 & 0 \end{bmatrix}, L_\epsilon = \begin{bmatrix} 0 & 0 & 0 & 1 \\ 0 & 0 & 1 & 0 \\ 0 & 1 & 0 & 0 \\ 1 & 0 & 0 & 0 \end{bmatrix}.$$

The symmetry of the matrices shows that the involution is trivial, when restricted to the basis $\mathcal{B}$. The dimension function is given by

$$d(1) = d(\epsilon) = 1, \ d(\alpha) = d(\beta) = \phi \ ,$$

where $\phi = \frac{1+\sqrt{5}}{2}$ denotes the so-called 'golden mean'.

(d) One reason for calling the example in (c) above the '$A_4$ fusion algebra' is that it is generated as an algebra by $\alpha$ and the matrix $L_\alpha$ is precisely the adjacency matrix of the Coxeter diagram $A_4$. There are fusion algebras which correspond similarly to each $A_n$; in fact, the same is true for each of the diagrams $E_6, E_8$ and $D_n$ provided $n$ is not of the form $(4k+1)$, while there exists no fusion algebra corresponding to $D_{4k+1}$ or $E_7$.

We next pause to discuss two 'classical' examples of fusion algebras.

EXAMPLE 1.3.5 Both these examples start from a fixed (but arbitrary) finite group $G$.

(a) Let $\mathcal{A} = \mathbb{C}G$ denote the complex group algebra; this has the basis $\mathcal{B} = G$, and the product and involution, when restricted to basis vectors, is just the product and inversion in the group. It is trivially verified that this gives a fusion algebra, and that the dimension function, when restricted to the basis, is identically equal to 1.

(b) Let $\hat{G}$ denote a set of irreducible unitary representations of $G$, which are pairwise inequivalent, and complete in the sense that every irreducible representation of $G$ is equivalent to one from the set at hand. Let $\mathcal{A}$ denote a vector space with basis $\hat{G}$. Then $\mathcal{A}$ is a fusion algebra, if we define $\pi^*$ to be the element of $\hat{G}$ which is equivalent to the 'conjugate' or 'contragredient' of the representation $\pi$, and if the structure constant $N^{\nu}_{\lambda,\mu}$ is defined to be the 'multiplicity with which $\nu$ occurs in $\lambda \otimes \mu$'. In this case, the dimension function is given by $d_\lambda = \dim V_\lambda$, where $V_\lambda$ is the vector space on which $\lambda$ represents $G$ as unitary operators.

(c) It must be noted that the values of the dimension function on the basis are all integral, in the case of these examples coming from a group or group-dual. In contrast, this is no longer true of the examples associated to the Coxeter diagrams (of Remark 1.3.2(d)). This is indicative of the reason for referring to the group and group-dual examples as 'classical' - the suggestion being that the other examples are 'quantum'.

DEFINITION 1.3.6 *A fusion algebra is said to have* **intertwiner spaces** *if, to every* $X, Y, Z \in \mathcal{B}_A$, *there is associated a finite-dimensional Hilbert space* $\mathcal{H}^Z_{XY}$ *such that:*

*(a)* $\dim \mathcal{H}^Z_{XY} = N^Z_{XY}$; *and*

*(b) there exist anti-unitary operators*

$$F_{(12)} : \mathcal{H}^Z_{XY} \to \mathcal{H}^X_{Z\bar{Y}} \ , \ F_{(23)} : \mathcal{H}^Z_{XY} \to \mathcal{H}^Y_{\bar{X}Z} \tag{1.3.22}$$

*such that*

$$F^2_{(12)} = F^2_{(13)} = (F_{(12)}F_{(23)})^3 = id \ . \tag{1.3.23}$$

Note that the compositions in the last equation are to be suitably interpreted. Thus, for example, the symbol $F^2_{(12)}$ denotes the composite map

$$\mathcal{H}^Z_{XY} \stackrel{F_{(12)}}{\to} \mathcal{H}^X_{Z\bar{Y}} \stackrel{F_{(12)}}{\to} \mathcal{H}^Z_{XY} \ ,$$

and it is required that this composite is the identity map of $\mathcal{H}^Z_{XY}$. The best way to think of these maps is as maps of $\oplus_{X,Y,Z}\mathcal{H}^Z_{XY}$ which map each summand to another appropriate summand. Since

$$S_3 =< x, y | x^2 = y^2 = (xy)^3 = 1 >$$

is a presentation of $S_3$, it follows from equation 1.3.23 that if we write $\mathcal{UA}(\mathcal{K})$ to denote the group whose typical element is a unitary or anti-unitary operator on the Hilbert space $\mathcal{K}$, then there exists a unique homomorphism

$$S_3 \ni \pi \mapsto F_\pi \in \mathcal{UA}(\oplus_{X,Y,Z}\mathcal{H}^Z_{XY})$$

such that $F_{(12)}$ and $F_{(23)}$ are the direct sums of all the operators denoted by the same symbol in equation 1.3.23. Thus, for instance, for any $X, Y, Z \in \mathcal{B}_A$, $F_{(123)} : \mathcal{H}_{XY}^{Z} \to \mathcal{H}_{Y\bar{Z}}^{\bar{X}}$ is, by definition, the composite map given by

$$\mathcal{H}_{XY}^{Z} \stackrel{F_{(23)}}{\to} \mathcal{H}_{\bar{X}Z}^{Y} \stackrel{F_{(12)}}{\to} \mathcal{H}_{Y\bar{Z}}^{\bar{X}}$$

and also equals the composite map given by

$$\mathcal{H}_{XY}^{Z} \stackrel{F_{(13)}}{\to} \mathcal{H}_{\bar{Y}\bar{X}}^{\bar{Z}} \stackrel{F_{(23)}}{\to} \mathcal{H}_{Y\bar{Z}}^{\bar{X}} .$$

Since $F_{(12)}$ and $F_{(23)}$ are antiunitary, it follows that $F_\pi$ is antiunitary if and only if $\pi$ is an odd permutation, i.e., $\pi$ is a transposition. Just for the sake of completeness, we note the following:

$$F_{(12)} : \mathcal{H}_{XY}^{Z} \to \mathcal{H}_{Z\bar{Y}}^{X}$$

$$F_{(23)} : \mathcal{H}_{XY}^{Z} \to \mathcal{H}_{\bar{X}Z}^{Y}$$

$$F_{(13)} : \mathcal{H}_{XY}^{Z} \to \mathcal{H}_{\bar{Y}\bar{X}}^{\bar{Z}}$$

$$F_{(123)} : \mathcal{H}_{XY}^{Z} \to \mathcal{H}_{Y\bar{Z}}^{\bar{X}}$$

$$F_{(132)} : \mathcal{H}_{XY}^{Z} \to \mathcal{H}_{\bar{Z}X}^{\bar{Y}}$$

The maps $F_\pi, \pi \in S_3$ will be referred to as the **Frobenius reciprocity maps**.

# Chapter 2

# The invariant

## 2.1 The construction

In this chapter, we construct of a (2+1)-dimensional TQFT by a *state sum* approach. We shall once and for all assume that we have a fusion algebra $\mathcal{A}$ with intertwiner spaces.

We shall first concentrate on the construction of the numerical invariant associated with a compact oriented 3-manifold. (We will later verify that what we have is really an invariant of the manifold.) We will start with a 3-manifold which comes presented with a triangulation (i.e., decomposition into tetrahedra, which have been glued along their faces). We shall compute the invariant as a 'state-sum', i.e., a sum of scalars, each one obtained by 'pairing the given triangulated 3-manifold with a state'. The first order of business is to clarify what states are; and before we go to the general 3-manifold, we shall identify what we mean by a state of a tetrahedron.

So suppose $T$ is a tetrahedron, whose 4 vertices we shall assume to be labelled as 1,2,3,4. Loosely speaking, a state $\phi$ does three things: (a) it provides a total order - call it $\leq_\phi$ - on the vertices of the tetrahedron, and consequently a 'direction' to each edge of the tetrahedron (whereby edges are directed from 'smaller' to 'larger' vertices); (b) it labels each edge, say $e = (ij)$, by an element, say $\phi^1(e)$, of $\mathcal{B}_{\mathcal{A}}$; and (c) for each face, say $f = (ijk)$ of the tetrahedron, where we may and do assume that $i <_\phi j <_\phi k$, it assigns an element $\phi^2(f)$ (which we shall call an intertwiner) of the Hilbert space $\mathcal{H}_\phi(f) = \mathcal{H}^{\phi^1(ik)}_{\phi^1(jk),\phi^1(ij)}$.

We formalise all this in the definition below.

DEFINITION 2.1.1 *A* **state** *of the tetrahedron is a triple of the form* $\phi = (\leq_\phi, \phi^1, \phi^2)$, *where:*

$\leq_\phi$ *is a total order on the set* $\{1, 2, 3, 4\}$ *(and hence induces an orientation on* $T$*);*

$\phi^1 : edge(T) \to \mathcal{B}_A$ *is a labelling of the edges of* $T$ *by members of the basis of the fusion algebra; and*

$\phi^2 : face(T) \to \oplus_{X,Y,Z}\mathcal{H}^Z_{X,Y}$ *is a labelling of the faces of* $T$ *by 'intertwiners' such that* $\phi^2(f) \in \mathcal{H}_\phi(f)$, *where* $\mathcal{H}_\phi(f)$ *is as above.*

*We shall denote the collection of all the states on the tetrahedron by* $\mathcal{S}(T)$.

We first 'classify' the states according to their total order. For this, we define

$$\mathcal{S}_\sigma(T) = \{\phi \in \mathcal{S}(T) : \sigma(1) <_\phi \sigma(2) <_\phi \sigma(3) <_\phi \sigma(4)\}\,,\ \forall\, \sigma \in S_4\,.$$

It should be evident that

$$\mathcal{S}(T) \;=\; \coprod_{\sigma \in S_4} \mathcal{S}_\sigma(T)\,,$$

and that

$$i <_\phi j \;\Leftrightarrow\; \sigma^{-1}(i) < \sigma^{-1}(j)\ \forall\ \phi \in \mathcal{S}_\sigma(T)$$

(since, by definition, $\sigma : (\{1,2,3,4\}, \leq) \to (\{1,2,3,4\}, \leq_\phi)$ is an order-isomorphism.)

In the sequel, we simply write $1$ for the identity element of $S_4$.

PROPOSITION 2.1.2 *(a) For each* $\sigma \in S_4$, *the following prescription defines a bijection*

$$\mathcal{S}_1(T) \ni \phi \;\mapsto\; \sigma \cdot \phi \;\in \mathcal{S}_\sigma(T)\ :$$

*(i) for* $1 \leq i, j \leq 4$, *define*

$$i <_{\sigma\cdot\phi} j \;\Leftrightarrow\; \sigma^{-1}(i) \leq \sigma^{-1}(j) \;;$$

*(ii) for* $e = (ij),\ 1 \leq i < j \leq 4$, *define*

$$(\sigma\cdot\phi)^1(e) \;=\; \begin{cases} \overline{\phi^1(e)} & \text{if } j <_{\sigma\cdot\phi} i \\ \phi^1(e) & \text{if } i <_{\sigma\cdot\phi} j \end{cases} \;;$$

*(iii) for* $f = (ijk),\ 1 \leq i < j < k \leq 4$, *define*

$$(\sigma\cdot\phi)^2(f) \;=\; F_\tau(\phi^2(f))$$

*where* $F_\tau$ *is as in the definition of intertwiner spaces, and* $\tau \in S_3$ *is defined thus:*

$$\begin{aligned} &\sigma^{-1}(m\text{-th smallest of } i,j,k) \\ &\quad = \; \tau(m)\text{-th smallest of } \sigma^{-1}(i), \sigma^{-1}(j), \sigma^{-1}(k) \;. \end{aligned}$$

*(b) The above prescription is a 'local' one in the following sense:*
*(i) for each edge 'e', the value of* $(\sigma\cdot\phi)^1(e)$ *depends only on* $\phi|_e$ *and* $\sigma$*; and*
*(ii) for each face 'f', the value of* $(\sigma\cdot\phi)^2(f)$ *depends only on* $\phi|_f$ *and* $\sigma$.
*(In (i) and (ii) above, we write* $\phi|_e$ *(resp.,* $\phi|_f$*) to mean the restrictions of* $\phi^1$ *to* $\{e\}$ *(resp., the data given by restricting* $\phi^1$ *to the bounding edges of* $f$ *and the value* $\phi^2(f)$*).)*

*(c) The following prescription defines a unique action of* $S_4$ *on* $\mathcal{S}(T)$*: if* $\pi \in S_4$ *and* $\psi \in \mathcal{S}(T)$*, then there exists a unique* $\sigma \in S_4$ *such that* $\psi \in \mathcal{S}_\sigma(T)$*; thus there exists a unique* $\phi \in \mathcal{S}_1(T)$ *such that* $\psi = \sigma\cdot\phi$*; define* $\pi\cdot\psi \;=\; (\pi\sigma)\cdot\phi$.

**Proof:** The proofs of (a) and (b) only involve straightforward verifications of the definitions. As for (c), it follows from (b) that any element of $\mathcal{S}(T)$ is uniquely expressible in the form $\sigma\cdot\phi$, for some $\sigma \in S_4$, $\phi \in \mathcal{S}_1(T)$; once this has been observed, (c) follows from associativity of group-multiplication. □

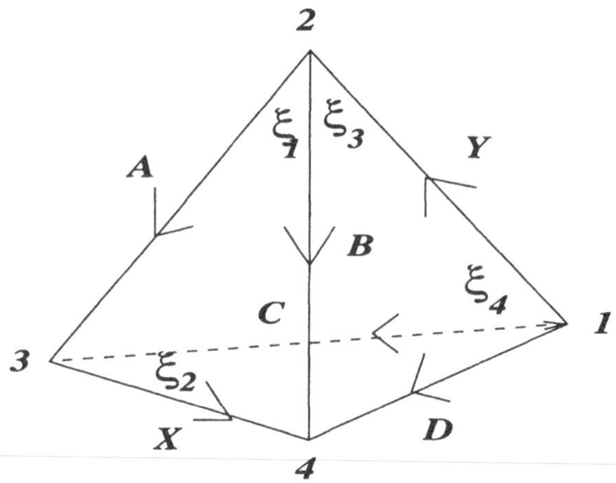


Figure 2.1: A state on the tetrahedron

REMARK 2.1.3 (a) The total order $\leq_\phi$ on the set of vertices of a tetrahedron $T$ (given by an element of $\mathcal{S}(T)$) endows an orientation on $T$. For a state $\phi \in \mathcal{S}_\sigma(T)$, we shall use the notation:

$$f_1(\phi) = (\sigma(2), \sigma(3), \sigma(4)) \ , \quad f_2(\phi) = (\sigma(1), \sigma(3), \sigma(4)) \ ,$$
$$f_3(\phi) = (\sigma(1), \sigma(2), \sigma(4)) \ , \quad f_4(\phi) = (\sigma(1), \sigma(2), \sigma(3)) \ ;$$

thus $f_i(\phi)$ is just the face opposite the $i$-th largest vertex; and we shall call $f_1(\phi)$ and $f_3(\phi)$ the **positive faces** of $(T, \leq_\phi)$ and $f_2(\phi)$ and $f_4(\phi)$ the **negative faces** of $(T, \leq_\phi)$.

(b) Define an equivalence relation $\sim_1$ on $\mathcal{S}(T)$ by demanding that $\phi \sim_1 \psi \Leftrightarrow (\leq_\phi, \phi_1) = (\leq_\psi, \psi_1)$. For each $\sim_1$-equivalence class, say $[\phi]_1$, note that the Hilbert space $\mathcal{H}_\phi(f_i(\phi))$ depends only on $[\phi]_1$; choose some orthonormal basis, say $\mathcal{B}_i([\phi]_1)$, for $\mathcal{H}_\phi(f_i(\phi))$. Given such a family $\mathcal{B} = \{\mathcal{B}_i([\phi]_1) : 1 \leq i \leq 4, \phi \in \mathcal{S}(T)\}$ of orthonormal bases, we shall write

$$\mathcal{S}(T; \mathcal{B}) \ = \ \{\phi \in \mathcal{S}(T) : \phi^2(f_i(\phi)) \in \mathcal{B}_i([\phi]_1) \ \forall \ 1 \leq i \leq 4 \ \} \ . \tag{2.1.1}$$

Figure 2.1 illustrates an element of $\mathcal{S}_1(T)$, and we have written $\xi_i$ for the intertwiner labelling the face $f_i$.

Some of the following considerations might become more transparent if the above picture is re-written in the alternative

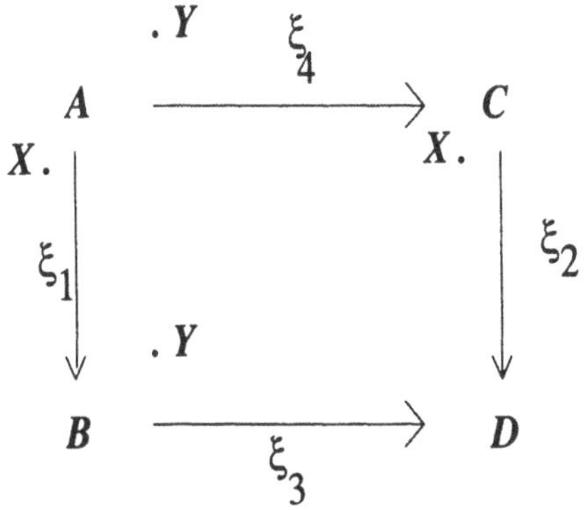


Figure 2.2: The 'dual picture' of a state

form shown in Figure 2.2, where a horizontal (resp., vertical) arrow from $P$ to $Q$ labelled by a symbol $\xi$ signifies that $\xi \in \mathcal{H}_{PY}^{Q}$ (resp., $\xi \in \mathcal{H}_{XP}^{Q}$).

We shall be discussing a *quantum 6j symbol*: this is an assignment of scalars to elements of $\mathcal{S}(T)$, which satisfies some conditions; the following way of thinking of states would make some of those conditions more transparent. In our application to subfactors, the members of $\mathcal{B}_A$ would be 'irreducible bimodules over some $II_1$ factor $N$', and the spaces $\mathcal{H}_{XY}^{Z}$ would denote precisely the collection of $N$-bilinear maps from $X \otimes_N Y$ to $Z$.

Then, given a rectangle as in Figure 2.2, we actually have the following diagram of $N$-bimodules and $N$-linear maps:

$$\begin{array}{ccc} X \otimes_N A \otimes_N Y & \stackrel{id_X \otimes \xi_4}{\to} & X \otimes_N C \\ \xi_1 \otimes id_Y \downarrow & & \downarrow \xi_2 \\ B \otimes_N Y & \stackrel{\xi_3}{\to} & D \end{array} \tag{2.1.2}$$

The above diagram is typically not a commutative diagram; however, it would yield the $N$-linear map

$$\xi_3(\xi_1 \otimes id_Y)(id_X \otimes \xi_4^*)\xi_2^* : D \to D$$

and this gives a scalar since the only $N$-linear maps of an irreducible bimodule are the scalar multiples of the identity operator.

DEFINITION 2.1.4 *A function*

$$\mathcal{S}(T) \ni \phi \overset{Z}{\mapsto} Z(\phi) \in \mathbb{C}$$

*is called a* **quantum 6j symbol** *if it satisfies the following conditions:*

*(i) (sesquilinearity) For any* $\phi_0 \in \mathcal{S}(T)$, *it should be clear that* $[\phi_0]_1 = \{\phi \in \mathcal{S}(T) : \phi \sim_1 \phi_0\}$ *(see Remark 2.1.3) is in natural bijection with* $\prod_{i=1}^4 \mathcal{H}_{\phi_0}(f_i(\phi_0))$. *Hence* $Z|_{[\phi_0]_1}$ *may be identified with a map* $Z : \prod_{i=1}^4 \mathcal{H}_{\phi_0}(f_i(\phi_0)) \to \mathbb{C}$. *This map is required to extend sesquilinearly to* $(\mathcal{H}_{\phi_0}(f_1(\phi_0)) \otimes_{\mathbb{C}} \mathcal{H}_{\phi_0}(f_3(\phi_0))) \times (\mathcal{H}_{\phi_0}(f_2(\phi_0)) \otimes_{\mathbb{C}} \mathcal{H}_{\phi_0}(f_4(\phi_0)))$; *in other words, it is required that the function* $Z(\xi_1, \xi_2, \xi_3, \xi_4)$ *depends linearly on* $\xi_1$ *and* $\xi_3$ *(the labels on the positive faces) and conjugate-linearly on* $\xi_2$ *and* $\xi_4$ *(the labels on the negative faces).*

*(ii) (tetrahedral symmetry) It is required that*

$$Z(\sigma \cdot \phi) = \begin{cases} \overline{Z(\phi)} & if\ \sigma \text{ is odd} \\ Z(\phi) & if\ \sigma \text{ is even} \end{cases}$$

*(iii) (unitarity) (For this condition, the reader should compare the pictorial depictions of states given by Figure 2.2 and the diagram (2.1.2).) Fix* $A, D, X, Y \in \mathcal{B}_A$. *Define a matrix* $U$ *by*

$$U(\ (\xi_1, B, \xi_3), (\xi_4, C, \xi_2)\ ) = \sqrt{d(B)d(C)}\ Z(\xi_1, \xi_2, \xi_3, \xi_4)\ ; \tag{2.1.3}$$

*here the rows of* $U$ *are indexed by triples* $(\xi_1, B, \xi_3)$ *where* $B \in \mathcal{B}_A, \xi_1 \in \mathcal{B}_1, \xi_3 \in \mathcal{B}_3$; *and the columns of* $U$ *are indexed by triples* $(\xi_4, C, \xi_2)$ *where* $C \in \mathcal{B}_A, \xi_4 \in \mathcal{B}_4, \xi_2 \in \mathcal{B}_2$; *here the symbols* $\mathcal{B}_1, \mathcal{B}_2, \mathcal{B}_3, \mathcal{B}_4$ *represent orthonormal bases of* $\mathcal{H}^B_{XA}, \mathcal{H}^D_{XC}, \mathcal{H}^D_{BY}$ *and* $\mathcal{H}^C_{AY}$ *respectively. This requirement for a quantum 6j symbol is that the matrix* $U$ *is unitary.*

*(iv) (2-3 condition) For any* $A, B, C, D, E, F, H, X, Y \in \mathcal{B}_A$ *and* $\xi_1 \in \mathcal{H}^B_{XA}, \xi_2 \in \mathcal{H}^D_{XC}, \xi_4 \in \mathcal{H}^C_{AY}, \eta_1 \in \mathcal{H}^H_{BE}, \eta_3 \in \mathcal{H}^D_{HF}, \eta_4 \in \mathcal{H}^Y_{EF}$, *we have*

$$\sum_{\xi_3} Z(\xi_1, \xi_2, \xi_3, \xi_4)\ Z(\eta_1, \xi_3, \eta_3, \eta_4) =$$

$$\sum_{G,\lambda,\mu,\nu} d(G) Z(\lambda, \xi_4, \mu, \eta_4) Z(\xi_1, \nu, \eta_1, \lambda) Z(\nu, \xi_2, \eta_3, \mu)\ ;$$

*here, the sum on the left ranges over* $\xi_3 \in \mathcal{B}_3$*, where* $\mathcal{B}_3$ *is an orthonormal basis for* $\mathcal{H}^D_{BY}$*; and the sum on the right ranges over* $G \in \mathcal{B}_\mathcal{A}$*,* $\lambda \in \mathcal{B}_5, \mu \in \mathcal{B}_6, \nu \in \mathcal{B}_7$ *where* $\mathcal{B}_5, \mathcal{B}_6$ *and* $\mathcal{B}_7$ *are orthonormal bases for* $\mathcal{H}^G_{AE}, \mathcal{H}^C_{GF}, \mathcal{H}^H_{XG}$ *respectively.*

REMARK 2.1.5 (a) The meaning of the (2-3) condition will become clear when we use it later.

(b) Note that the matrix $U$ defined in the unitarity requirement in the above definition has rows and columns indexed by different sets; the number of rows of $U$ is $\sum_B N^B_{XA} N^D_{BY}$ , while the number of columns is $\sum_C N^C_{AY} N^D_{XC}$ , and both these sums are equal since they are both equal to $\langle XAY, D\rangle$ (in $\mathcal{A}$).

(c) It is true that requiring unitarity with respect to one choice of bases $\{\mathcal{B}_i\}_{i=1}^4$ of the underlying Hilbert spaces implies unitarity with respect to any choice of bases. This is a fairly straightforward consequence of the sesquilinearity axiom.

Before proceeding further, we wish to draw a consequence of the tetrahedral symmetry and unitarity requirements on a quantum 6j symbol, which we shall refer to as biunitarity.

PROPOSITION 2.1.6 *(biunitarity) If* $Z$ *is a quantum 6j symbol, and if* $B, C, X, Y \in \mathcal{B}_\mathcal{A}$ *are fixed, the following matrix* $V$ *is unitary:*

$$V(\ (\xi_2, D, \xi_3), (\xi_4, A, \xi_1)\ ) \ = \ \sqrt{d(A)d(D)}\ \overline{Z(\xi_1, \xi_2, \xi_3, \xi_4)}\ ;$$

*here, the rows of the matrix* $V$ *are indexed by triples* $(\xi_2, D, \xi_3)$*, where* $D \in \mathcal{B}_\mathcal{A}$ *and* $\xi_2 \in \mathcal{B}'_2, \xi_3 \in \mathcal{B}'_3$ *and* $\mathcal{B}'_2, \mathcal{B}'_3$ *are orthonormal bases for* $\mathcal{H}^D_{XC}, \mathcal{H}^D_{BY}$*, respectively; and the columns are indexed by triples* $(\xi_4, A, \xi_1)$*, where* $A \in \mathcal{B}_\mathcal{A}, \xi_4 \in \mathcal{B}'_4, \xi_1 \in \mathcal{B}'_1$*, and* $\mathcal{B}'_4, \mathcal{B}'_1$ *are orthonormal bases for* $\mathcal{H}^C_{AY}, \mathcal{H}^X_{AB}$ *respectively.*

**Proof:** It will help to think thus about the unitarity axiom: fix $\sigma \in S_4$ and $A', D', X', Y' \in \mathcal{B}_\mathcal{A}$ and suppose $\phi_0 \in \mathcal{S}_\sigma(T)$ is $Y' = \phi_0^1(\{\sigma(1), \sigma(2)\}), A' = \phi_0^1(\{\sigma(2), \sigma(3)\}), X' = \phi_0^1(\{\sigma(3), \sigma(4)\}), D' = \phi_0^1(\{\sigma(4), \sigma(1)\})$. Suppose, further, that we have chosen some collection $\mathcal{B} = \{\mathcal{B}_i([\phi]_1)\}$ as in equation

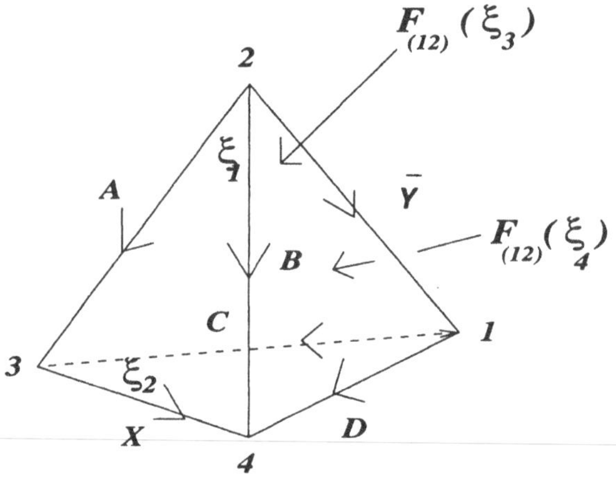


Figure 2.3: The Frobenius map of a state

(2.1.1). Then the unitarity axiom states that the function defined on $[\phi_0]_1 \cap \mathcal{S}(T;\mathcal{B})$ by

$$\phi \mapsto \sqrt{d(\phi^1(\{\sigma(2),\sigma(4)\}))d(\phi^1(\{\sigma(1),\sigma(3)\}))}\; Z(\phi)$$

yields a unitary matrix with rows and columns indexed by the ordered triples given by $(\phi^2(f_1(\phi)), \phi^1(\{\sigma(2),\sigma(4)\}), \phi^2(f_3(\phi)))$ and $(\phi^2(f_4(\phi)), \phi^1(\{\sigma(1),\sigma(3)\}), \phi^2(f_2(\phi)))$.

Now, let $\phi \in \mathcal{S}_1(T)$ be the state illustrated in Figure 2.1. It is then seen that $(12)\cdot\phi \in \mathcal{S}_{(12)}(T)$ is the state illustrated in Figure 2.3.

Define the bases $\mathcal{B}'_i$ for $\mathcal{H}_{(12)\cdot\phi}(f_i((12)\cdot\phi))$, in terms of the bases $\mathcal{B}_i$ chosen for the definition of the '$U$-matrix', thus:

$$\mathcal{B}'_1((12)\cdot\phi) = \mathcal{B}_2(\phi)\ ,\ \mathcal{B}'_3((12)\cdot\phi) = F_{(12)}(\mathcal{B}_3(\phi)),$$

$$\mathcal{B}'_2((12)\cdot\phi) = \mathcal{B}_1(\phi)\ ,\ \mathcal{B}'_4((12)\cdot\phi) = F_{(12)}(\mathcal{B}_4(\phi))\ .$$

It is then seen, from the reformulation of unitarity given in the first paragraph of this proof, that if we define

$$V((\xi_2,D,\xi_3),(\xi_4,A,\xi_1)) = U((\xi_2,D,F_{(12)}(\xi_3)),(F_{(12)}(\xi_4),A,\xi_1)),$$

then we have

$$V((\xi_2,D,\xi_3),(\xi_4,A,\xi_1)) \;=\; U((\xi_2,D,F_{(12)}(\xi_3)),(F_{(12)}(\xi_4),A,\xi_1))$$

$$\begin{aligned} &= \sqrt{d(D)d(A)}Z((12)\cdot\phi) \\ &= \sqrt{d(D)d(A)}\ \overline{Z(\phi)} \\ &= \sqrt{d(D)d(A)}\ \overline{Z(\xi_1,\xi_2,\xi_3,\xi_4)} \end{aligned}$$

and the proof is complete. □

We are now ready to pass from tetrahedra to general closed oriented 3-manifolds. Suppose $M$ is such a manifold, and suppose $\Delta$ is a triangulation (= simplicial decomposition) of $M$. We shall write $T \in \Delta$ to signify that $T$ is a 3-simplex (=tetrahedron) occurring in $\Delta$.

DEFINITION 2.1.7 *Suppose '$\leq$' is a total order on the set of vertices of $\Delta$. By a state $\phi$ on $\Delta$ that is compatible with $\leq$, we shall mean a family $\phi = \{\phi_T : T \in \Delta\}$ such that, for every $S, T \in \Delta$, we have: (i) $\phi_T \in \mathcal{S}(T)$; (ii) $\leq_{\phi_T} = \leq|_{Vert(T)}$; and (iii) if $S$ and $T$ meet in a non-empty-face $f$, then $(\phi_T)|_f = (\phi_S)|_f$.*

*We shall denote the set of such $\phi$'s by $\mathcal{S}(\Delta;\leq)$.*

REMARK 2.1.8 (a) Note that what we have called a 'state on $\Delta$ that is compatible with $\leq$' just amounts to a specification of an element $\phi^1(e) \in \mathcal{B}_A$ for every edge (=1-simplex) of $\Delta$, and of an intertwiner $\phi^2(f)$ for each face (=2-simplex) of $\Delta$, such that the restriction of this specification to any 3-simplex $T$ defines what we earlier called a state $\phi_T$ of the tetrahedron $T$ such that condition (ii) of the above definition is satisfied.

(b) As in remark 2.1.3(a), we shall define an equivalence relation $\sim_1$ on $\mathcal{S}(\Delta;\leq)$ by setting $\phi \sim_1 \psi$ if $\phi^1(e) = \psi^1(e)$ for every edge $e$ in $\Delta$.

Also, by condition (iii) of Definition 2.1.7, we see that the Hilbert space $\mathcal{H}_f(\phi)$ is unambiguously defined for every face of $\Delta$; clearly this Hilbert space depends only on $[\phi]_1$, and thus $\mathcal{H}_f([\phi]_1)$ makes sense.

As in remark 2.1.3(b), suppose we have chosen some collection $\mathcal{B} = \{\mathcal{B}_f([\phi]_1) : f \text{ a 2-simplex of } \Delta, \phi \in \mathcal{S}(\Delta;\leq)\}$ of orthonormal bases; then we shall write

$$\mathcal{S}(\Delta;\leq;\mathcal{B}) = \{\phi \in \mathcal{S}(\Delta;\leq) : \phi^2(f) \in \mathcal{B}_f([\phi]_1)\ \forall f\} .$$

Temporarily fix a state $\phi \in \mathcal{S}(\Delta; \leq)$. If $T \in \Delta$, then $T$ acquires an orientation - call it $or_{\leq}(T)$ - from the order $\leq_{\phi_T}$; this may or may not be the same as the orientation - call it $or_M(T)$ - that $T$ acquires as a submanifold of $M$. Define

$$\tilde{Z}(\phi_T) = \begin{cases} \overline{Z(\phi_T)} & if\ or_{\leq}(T) \neq or_M(T) \\ Z(\phi_T) & if\ or_{\leq}(T) = or_M(T) \end{cases} . \tag{2.1.4}$$

Also, let

$$< \phi > = \prod_{T \in \Delta} \tilde{Z}(\phi_T) ,$$

$$d(\phi) = \prod_{e \in \Delta} \phi^1(e) ,$$

and finally define

$$< (M, \Delta, \leq, \mathcal{B}) > = w^{-|Vert(\Delta)|} \sum_{\phi \in \mathcal{S}(\Delta; \leq; \mathcal{B})} d(\phi) < \phi > . \tag{2.1.5}$$

## 2.2 Invariance

We shall now make a modest start towards our goal of establishing that $< (M, \Delta, \leq, \mathcal{B}) >$ depends only upon the oriented homeomorphism class of the manifold $M$.

LEMMA 2.2.1 *(a)* $< (M, \Delta, \leq, \mathcal{B}) > = < (M, \Delta, \leq, \mathcal{B}') >$, *for any two choices* $\mathcal{B}, \mathcal{B}'$ *of bases.*

*(b)* $< (M, \Delta, \leq, \mathcal{B}) > = < (M, \Delta, \leq', \mathcal{B}) >$, *for any two total orders* $\leq, \leq'$.

**Proof:** (a) It clearly suffices to establish the result in the case when there exists a $\phi_0 \in \mathcal{S}(\Delta, \leq \mathcal{B})$ and a face $f_0$ of $\Delta$ such that $\mathcal{B}_f([\phi]_1) = \mathcal{B}'_f([\phi]_1)$ unless $\phi \sim_1 \phi_0$ and $f = f_0$ (i.e., the case when the two choices differ only at one place).

Consider the equivalence relation $\sim_0$ defined on $[\phi_0]_1$ by requiring that $\phi \sim_0 \psi$ if and only if $\phi^2(f) = \psi^2(f)\ \forall\ f \neq f_0$. For each equivalence class $[\psi]_0$ in $[\phi_0]_1$, let $\mathcal{K}_{[\psi]_0} = \mathcal{S}(\Delta, \leq) \cap [\psi]_0$, and note that there is a natural bijection

$$\mathcal{K}_{[\psi]_0} \ni \phi \overset{F}{\mapsto} \phi(f_0) \in \mathcal{H}_{f_0}([\phi_0]) \ .$$

Since $M$ is a closed 3-manifold, there exist precisely two tetrahedra, say $T_1, T_2 \in \Delta$ which contain $f_0$ as a face; further, we may assume the labelling is so that $f_0$ is a positive (resp., negative) face of $T_1$ (resp., $T_2$) (with respect to their orientations as submanifolds of $M$). It then follows from the 'sesquilinearity' condition on quantum 6j symbols that there exist linear functionals $\alpha_1, \alpha_2$ on $\mathcal{H}_{f_0}([\phi_0])$ such that $\tilde{Z}(\phi_{T_1}) = \alpha_1(F(\phi))$ and $\tilde{Z}(\phi_{T_2}) = \overline{\alpha_2(F(\phi))}$ for all $\phi \in \mathcal{K}_{[\psi]_0}$. The definitions imply that $\prod_{T\in\Delta} \tilde{Z}(\phi_T) = C_{[\psi]_0}\alpha_1(F(\phi))\overline{\alpha_2(F(\phi))}$ for all $\phi \in \mathcal{K}_{[\psi]_0}$, where $C_{[\psi]_0} = \prod_{T\in\Delta, T\neq T_1,T_2} \tilde{Z}(\psi_T)$.

Now there exist unique vectors $\zeta_i \in \mathcal{H}_{f_0}([\phi_0])$ such that $\alpha_i(\cdot) = \langle \cdot, \zeta_i \rangle$, for $i = 1, 2$. Hence if $\{\xi_j : 1 \leq j \leq n\}$ is any orthonormal basis for $\mathcal{H}_{f_0}([\phi_0])$, note that

$$\begin{aligned} \sum_{j=1}^{n} \alpha_1(\xi_j)\overline{\alpha_2(\xi_j)} &= \sum_{j=1}^{n} \langle \xi_j, \zeta_1 \rangle \langle \zeta_2, \xi_j \rangle \\ &= \langle \zeta_2, \zeta_1 \rangle \ . \end{aligned} \tag{2.2.6}$$

Now write $< (M, \Delta, \leq, \mathcal{B}) >= S_1 + \Sigma_{[\psi]_0} S_{[\psi]_0}$, where $S_1$ (resp., $S_{[\psi]_0}$) denotes the subsum, of the right side of equation (2.1.5), over those $\phi \in \mathcal{S}(\Delta, \leq \mathcal{B})$ which also satisfy $\phi \notin [\phi_0]_1$ (resp., $\phi \in \mathcal{K}_{[\psi]_0}$). We also have a splitting $< (M, \Delta, \leq, \mathcal{B}') >= S'_1 + \Sigma_{[\psi]_0} S'_{[\psi]_0}$. Then note that $S_1 = S'_1$ (term by term), and that

$$\begin{aligned} S_{[\psi]_0} &= \sum_{\xi\in\mathcal{B}_{f_0}([\phi_0])} C_{[\psi]_0}\alpha_1(\xi)\overline{\alpha_2(\xi)} \\ &= C_{[\psi]_0}\langle \zeta_2, \zeta_1 \rangle \ ; \end{aligned}$$

we have an identical computation for $S'_{[\psi]_0}$, and we finally deduce that indeed $< (M, \Delta, \leq, \mathcal{B}) > \ - \ < (M, \Delta, \leq, \mathcal{B}') >$.

(b) If $\leq, \leq'$ are two total orders, then we shall set up a bijection $\mathcal{S}(\Delta, \leq, \mathcal{B}) \ni \phi \mapsto \phi' \in \mathcal{S}(\Delta, \leq', \mathcal{B})$ such that $d(\phi) = d(\phi')$

and $\tilde{Z}(\phi_T) = \tilde{Z}(\phi'_T)$ for every $T \in \Delta$. For this, suppose $\phi = ((\phi_T))_{T\in\Delta} \in \mathcal{S}(\Delta, \leq, \mathcal{B})$. Fix $T \in \Delta$ and note that $\leq$ and $\leq'$ induce two total orders on the set of vertices of $T$; so there exists a unique permutation $\sigma \in S_4$ which will 'map $\leq$ to $\leq'$'. Define $\phi'_T = \sigma \cdot \phi_T$. The fact that the $S_4$ action on $\mathcal{S}(T)$ is a 'local one' - see Proposition 2.1.2(b) - implies that $\{\phi'_T : T \in \Delta\}$ 'patch up' to define an element $\phi' \in \mathcal{S}(\Delta, \leq')$. The fact that $d(X) = d(\bar{X})$ implies that $d(\phi) = d(\phi')$, while the fact that $\tilde{Z}(\phi_T) = \tilde{Z}(\phi'_T)$ is a consequence of the tetrahedral symmetry of the quantum 6j symbol.

□

Hence we may write $< (M, \Delta) >$ for what we have been calling $< (M, \Delta, \leq, \mathcal{B}) >$, and feel free to use any total order $\leq$ and choice $\mathcal{B}$ of bases when computing $< (M, \Delta) >$. Our next task is to show that $< (M, \Delta) >$ is actually independent of the triangulation $\Delta$. For this, we shall rely on the following result due to Pachner - see [Pac].

THEOREM 2.2.2 *(Pachner) If $\Delta$ and $\Delta'$ are triangulations of a smooth closed 3-manifold, then it is possible to pass from $\Delta$ to $\Delta'$ by a finite sequence of one of the following types of moves:*

*(isotopy) There exists an ambient isotopy which carries $\Delta$ to $\Delta'$ (i.e., there exists a continuous map $f : M \times [0, 1] \to M$ such that: (i) $x \mapsto f(x, t)$ is a diffeomorphism of $M$, for all $t \in [0, 1]$; (ii) $f(x, 0) = x \ \forall x \in M$; and (iii) $f(\Delta \times \{1\}) = \Delta'$).*

*(1-4 move) $\{\Delta, \Delta'\} = \{\Delta_1, \Delta_2\}$, and there exists a tetrahedron $T \in \Delta_1$ such that if $\{1, 2, 3, 4\}$ is the set of vertices of $T$, there exists a point 5 in the interior of $T$ such that $\Delta_2 = (\Delta_1 \backslash \{T\}) \cup \{T_i : 1 \leq i \leq 4\}$, where $T_i$ is the tetrahedron with vertices $(\{1, 2, 3, 4\} \backslash \{i\}) \cup \{5\}$.*

*(2-3 move) $\{\Delta, \Delta'\} = \{\Delta_1, \Delta_2\}$ and there exists two tetrahedra $T, T' \in \Delta_1$ which share a face $f$ in common such that if the vertex sets of $f, T$, and $T'$ are given by $\{1, 2, 4\}, \{1, 2, 3, 4\}$ and $\{1, 2, 3', 4\}$, respectively, then $\Delta_2 = \{T_1, T_2, T_4\} \cup (\Delta_1 \backslash \{T, T'\})$, where $T_i$ is the tetrahedron with vertex set $(\{1, 2, 4\} \backslash \{i\}) \cup \{3, 3'\}$ for $i = 1, 2, 4$.* □

Hence, in order to prove that $< (M, \Delta) >$ is independent and hence an invariant of the manifold, we just need to verify that

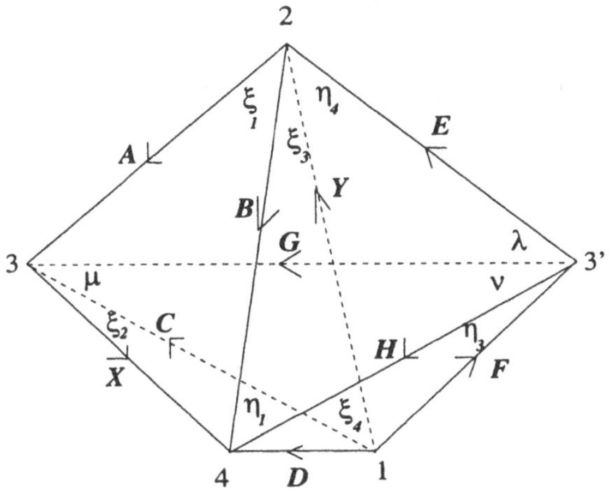


Figure 2.4: The tetrahedra for the 2-3 move

it remains invariant under each of the foregoing three kinds of moves. The fact that we have invariance under ambient isotopy follows from the construction (and the fact that an ambient isotopy preserves orientation).

We next verify invariance under what we called a 2-3 move. So consider the labelled tetrahedra in Figure 2.4.

In the language of what we called the (2-3) type move in Pachner's theorem, $T$ is the tetrahedron with vertices 1,2,3,4 in Figure 2.4, while $T'$ is the tetrahedron with vertices 1,3',2,4, $T_1$ has vertices 3',2,3,4, $T_2$ has vertices 1,3',3,4, and $T_4$ has vertices 1,3',2,3. Suppose that $\Delta$ is a triangulation of a closed oriented 3-manifold $M$ which contains $T, T'$, and that $\Delta'$ is the triangulation obtained by replacing $\{T, T'\}$ by $\{T_1, T_2, T_4\}$, but retaining the set $\mathcal{T}$ of all the remaining tetrahedra.

We also suppose that we have chosen a total order $\leq$ on the set of vertices of $\Delta$ such that: (i) $1 < 3' < 2 < 3 < 4$; and (ii) the orientation of $T$ that comes from the above ordering of its vertices is the same as the orientation it inherits as a submanifold of $M$. Given a state $\phi \in \mathcal{S}(\Delta, \leq)$ and a state $\phi' \in \mathcal{S}(\Delta', \leq)$, both have to be specified on the edges and faces of the tetrahedra of $\mathcal{T}$, as well as on the edges and faces of $T \cup T'$ that we have labelled as $A, B, C, D, E, F, H, X, Y$ and

$\xi_1, \xi_2, \xi_4, \eta_1, \eta_3, \eta_4$. Where they differ is that, in addition, $\phi$ must also label the face labelled $\xi_3$ above, while $\phi'$ must label the edge labelled $G$ as well as the faces labelled $\lambda, \mu, \nu$. The fact that we have $< (M, \Delta) > \;=\; < (M, \Delta') >$ now follows from the definitions and what we have called the (2-3) condition that has to be satisfied by quantum 6j symbols.

We now pass to invariance under moves of type (1-4). We assume, as in Theorem 2.2.2, that $\Delta = \mathcal{T} \cup \{T\}$, where $T$ is a tetrahedron with vertices $\{1, 2, 3, 4\}$; we also assume that $\leq$ is a total order on the vertices of $\Delta$ such that $1 < 2 < 3 < 4$. Suppose now that we choose and fix a point in the interior of $T$, which we shall denote by 2.5 (rather than by 5, as in the statement of Theorem 2.2.2) for reasons that will be obvious in a moment. For $1 \leq i \leq 4$, we consider a tetrahedron $T_i$ with vertex set $\{1, 2, 2.5, 3, 4\} \backslash \{i\}$, and consider the triangulation $\Delta'$ of $M$ given by $\Delta' = (\Delta \backslash \{T\}) \cup \{T_i : 1 \leq i \leq 4\}$, and the total order $\leq'$ on the vertices of $\Delta'$ defined by requiring that $2 <' 2.5 <' 3$ and that $x <' y \Leftrightarrow x < y$ for vertices $x, y$ of $\Delta$.

Arguing as in the case of (2-3) moves, we can see that the desired equation $< (M, \Delta, \leq) > \;=\; < (M, \Delta', \leq') >$ will follow if we have been able to verify the following condition:

If $A, B, C, D, X, Y \in \mathcal{B}_{\mathcal{A}}$ and $\xi_1 \in \mathcal{H}^B_{XA}, \xi_2 \in \mathcal{H}^D_{XC}, \xi_3 \in \mathcal{H}^D_{BY}$, $\xi_4 \in \mathcal{H}^C_{AY}$, we have:

$$\begin{aligned} w\; & Z(\xi_1, \xi_2, \xi_3, \xi_4) \\ = & \sum_{\{Z_i\},\{\xi_{ij}\}} \left( \prod_{i=1}^{4} d(Z_i) \right) \overline{Z(\xi_{12}, \xi_1, \xi_{13}, \xi_{14}) Z(\xi_{14}, \xi_{24}, \xi_4, \xi_{34})} \\ & \times \quad Z(\xi_{12}, \xi_2, \xi_{23}, \xi_{24}) Z(\xi_{13}, \xi_{23}, \xi_3, \xi_{34}) \,, \qquad (2.2.7) \end{aligned}$$

where the sum on the left is over $Z_1, Z_2, Z_3, Z_4 \in \mathcal{B}_{\mathcal{A}}$ and $\xi_{ij} \in \mathcal{B}_{ij}$ where $\mathcal{B}_{12}, \mathcal{B}_{13}, \mathcal{B}_{14}, \mathcal{B}_{23}, \mathcal{B}_{24}, \mathcal{B}_{34}$ are orthonormal bases for the intertwiner spaces $\mathcal{H}^{Z_4}_{XZ_3}, \mathcal{H}^B_{Z_4 Z_2}, \mathcal{H}^A_{Z_3 Z_2}, \mathcal{H}^D_{Z_4 Z_1}, \mathcal{H}^C_{Z_3 Z_1}, \mathcal{H}^{Z_1}_{Z_2 Y}$ respectively.

We begin by applying the (2-3) condition to write

$$\begin{aligned} & \sum_{\xi_{23}} Z(\xi_{12}, \xi_2, \xi_{23}, \xi_{24}) Z(\xi_{13}, \xi_{23}, \xi_3, \xi_{34}) \\ & \quad = \sum_{G,\lambda,\mu,\nu} d(G) Z(\lambda, \xi_{24}, \mu, \xi_{34}) Z(\xi_{12}, \nu, \xi_{13}, \lambda) Z(\nu, \xi_2, \xi_3, \mu) \end{aligned}$$

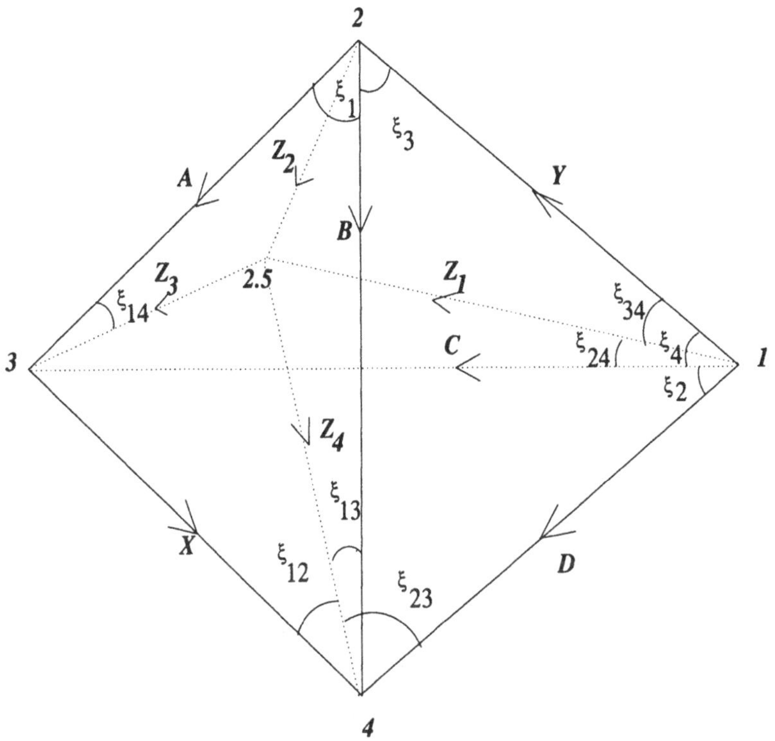


Figure 2.5: The tetrahedra for a 1-4 move

and substitute this expression to re-write the right side of equation (2.2.7) as

$$
\begin{aligned}
RHS & \\
&= \sum \left(\prod_{i=1}^{4} d(Z_i)\right) \overline{Z(\xi_{12},\xi_1,\xi_{13},\xi_{14}) Z(\xi_{14},\xi_{24},\xi_4,\xi_{34})} \\
&\quad \times \left(\sum d(G) Z(\lambda,\xi_{24},\mu,\xi_{34}) Z(\xi_{12},\nu,\xi_{13},\lambda) Z(\nu,\xi_2,\xi_3,\mu)\right)
\end{aligned}
$$

where the first sum above ranges over $Z_1, Z_2, Z_3, Z_4 \in \mathcal{B}_{\mathcal{A}}$ and over $\{\xi_{ij} : (i,j) \neq (2,3)\}$; and the second sum ranges over $\{G, \lambda, \mu, \nu\}$.

Now invoke the unitarity condition on the quantum 6j symbols to note that

$$
\begin{aligned}
\sum_{\xi_{12}, Z_4, \xi_{13}} \sqrt{d(A)d(G)}\, d(Z_4) \overline{Z(\xi_{12},\xi_1,\xi_{13},\xi_{14})} Z(\xi_{12},\nu,\xi_{13},\lambda) & \\
= \delta_{(\xi_1, A, \xi_{14}),(\nu, G, \lambda)} \ ;
\end{aligned}
$$

substitute this and re-write the previous equation as:

$$
\begin{aligned}
RHS &= \sum_{Z_1,Z_2,Z_3} \sum_{\xi_{14},\xi_{24},\xi_{34},\mu} (d(Z_1)d(Z_2)d(Z_3)) \times \\
&\quad \times \overline{Z(\xi_{14},\xi_{24},\xi_4,\xi_{34})}\ Z(\xi_{14},\xi_{24},\mu,\xi_{34}) Z(\xi_1,\xi_2,\xi_3,\mu) \ .
\end{aligned}
$$

Appeal again to the unitarity condition to note that

$$
\begin{aligned}
\sum_{\xi_{24}, Z_1, \xi_{34}} d(Z_1)d(A)\ \overline{Z(\xi_{14},\xi_{24},\xi_4,\xi_{34})}\ Z(\xi_{14},\xi_{24},\mu,\xi_{34}) & \\
= \delta_{(\xi_{14}, A, \xi_4),(\xi_{14}, A, \mu)} \ ,
\end{aligned}
$$

and hence that

$$
\sum_{\xi_{24}, Z_1, \xi_{34}} d(Z_1)\ \overline{Z(\xi_{14},\xi_{24},\xi_4,\xi_{34})}\ Z(\xi_{14},\xi_{24},\mu,\xi_{34}) = \frac{1}{d(A)}\ \delta_{\xi_4,\mu} \ .
$$

Hence deduce that

$$
RHS = \sum_{Z_2, Z_3, \xi_{14}, \mu} d(Z_2)d(Z_3)\ Z(\xi_1,\xi_2,\xi_3,\mu)\ \frac{1}{d(A)}\ \delta_{\xi_4,\mu}
$$

$$\begin{aligned}
&= \sum_{Z_2,Z_3,\xi_{14}} \frac{d(Z_2)d(Z_3)}{d(A)} Z(\xi_1,\xi_2,\xi_3,\xi_4) \\
&= \frac{Z(\xi_1,\xi_2,\xi_3,\xi_4)}{d(A)} \sum_{Z_2,Z_3} N^A_{Z_3Z_2} d(Z_2)d(Z_3) \\
&= w\, Z(\xi_1,\xi_2,\xi_3,\xi_4)\ ,
\end{aligned}$$

where we have used equation (1.3.21) in the last step; the proof of the verification of invariance under moves of type (1-4), and hence under choice of the underlying triangulation, is complete. Hence the quantity $<(M,\Delta,\leq,\mathcal{B})>$ is actually an invariant of the oriented manifold $M$.

We now proceed to the description of the TQFT coming from this construction.*

We begin with the description of the Hilbert space $V(\Sigma)$ associated with a closed oriented surface $\Sigma$. This will be defined as an appropriate subspace of a Hilbert space $W(\Sigma)$, which we shall first describe. Suppose, to start with, that $\Sigma$ is equipped with a triangulation $\delta$, and that the vertex set of $\delta$ has been equipped with a total order $\leq$. We shall define a *surface state* to be a pair $\phi = (\phi^1,\phi^2)$, where $\phi^1$ is a labelling of the edges of $\delta$ by elements of $\mathcal{B}_A$, and $\phi^2$ is a function which assigns, to each face $f$ of $\delta$, an element of the intertwiner space $\mathcal{H}_\phi(f)$ (where, as before, if $f=\{i,j,k\}$, with $i<j<k$, we write $\mathcal{H}_\phi(f) = \mathcal{H}^{\phi^1(ik)}_{\phi^1(jk),\phi^1(ij)}$).

Let $\mathcal{S}(\delta,\leq)$ denote the set of such surface states which are compatible with the order $\leq$ as above. As in the case of closed 3-manifolds, we define an equivalence relation $\sim_1$ on $\mathcal{S}(\delta,\leq)$ by requiring that $\phi \sim_1 \psi \Leftrightarrow \phi^1 = \psi^1$.

Also define $\tilde{\mathcal{H}}_\phi(f)$ to be equal to $\mathcal{H}_\phi(f)$ or $\overline{\mathcal{H}_\phi(f)}$ according as the orientation that $f$ gets from the order $\leq$ agrees with or differs from the orientation that $f$ gets as a submanifold of $\Sigma$. It should be noted, as in the earlier case, that $\tilde{\mathcal{H}}_\phi(f)$ depends only on the $\sim_1$-equivalence-class $[\phi]_1$ of $\phi$.

As a first approximation, we define

$$W(\Sigma,\delta,\leq) \;=\; \oplus_{[\phi_1]} \;\otimes_{f\in\delta}\, \tilde{\mathcal{H}}_{\phi_1}(f)\ . \qquad (2.2.8)$$

*It is not known, at this time, if this q.f. is cobordism-generated.

It should be noted that each $\phi \in \mathcal{S}(\delta, \leq)$ defines a vector $\xi_\phi$ living in the $[\phi]_1$-th direct summand, which is the 'product tensor' with $f$-th component equal to $\widetilde{\phi^2(f)}$ where we write $\tilde{\xi}$ to denote $\xi$ or $\bar{\xi}$ according as $\tilde{\mathcal{H}}$ is $\mathcal{H}$ or $\bar{\mathcal{H}}$. It should be evident that $\{\xi_\phi : \phi \in \mathcal{S}(\delta, \leq)\}$ is a total set in $W(\Sigma, \delta, \leq)$.

As in the case of closed 3-manifolds, we may use the Frobenius reciprocity maps $\{F_\pi : \pi \in S_3\}$ to deduce the existence of a naturally defined unitary operator

$$U_{\leq, \leq'} : W(\Sigma, \delta, \leq) \to W(\Sigma, \delta, \leq') \tag{2.2.9}$$

for any two total orders $\leq, \leq'$ on the vertex set of $\delta$. (In fact one uses the $F_\pi$'s to establish a bijection

$$\{\xi_\phi : \phi \in \mathcal{S}(\delta, \leq)\} \leftrightarrow \{\xi_{\phi'} : \phi' \in \mathcal{S}(\delta, \leq')\} \tag{2.2.10}$$

such that $U_{\leq, \leq'} \xi_\phi = \xi_{\phi'} \; \forall \phi \in \mathcal{S}(\delta, \leq)$.)

We now wish to generalise our construction of the invariant (for closed 3-manifolds) to 3-manifolds with boundary. Thus, we now suppose that $M$ is a (compact oriented) 3-manifold with $\partial M = \Sigma$. Suppose, as above, that $\delta$ is a triangulation of $\Sigma$ and that $\leq$ is a total order on the vertex set of $\delta$. Choose and fix (i) any triangulation $\Delta$ of $M$ such that $\Delta \cap \Sigma = \delta$, and (ii) any total order $\leq_M$ on the vertices of $\Delta$ such that the restriction of $\leq_M$ to the vertices of $\delta$ is $\leq$. We then define a state of $(M, \Delta, \leq_M)$ to be a pair $\Phi = (\Phi^1, \Phi^2)$, where $\Phi^1$ is a labelling of the edges of $\Delta$ by members of $\mathcal{B}_A$, and $\Phi^2$ is a labelling of each face (= 2-simplex) of $\Delta$ by an appropriate intertwiner. (Equivalently, if, for each tetrahedron $T \in \Delta$, we define $\Phi_T = (\leq_M |_{Vert(T)}, \Phi^1|_{Edge(T)}, \Phi^2|_{Face(T)})$, then $\Phi_T \in \mathcal{S}(T)$.) As before, we write $\mathcal{S}(M, \Delta, \leq_M)$ for the collection of such states. Next, we consider the equivalence relation $\sim_1$ on $\mathcal{S}(M, \Delta, \leq_M)$ defined by $\Phi \sim_1 \Psi \Leftrightarrow \Phi^1 = \Psi^1$, and note that the Hilbert space $\mathcal{H}_\Phi(f)$ depends only on $[\Phi]_1$. As before - see Remark 2.1.8 - we will be interested in making choices $\mathcal{B} = \{\mathcal{B}_f([\Psi]_1) : \Psi \in \mathcal{S}(M, \Delta, \leq_M), f \in Face(\Delta) \backslash Face(\delta)\}$ of orthonormal bases and considering the restricted collection $\mathcal{S}(M, \Delta, \leq_M, \mathcal{B})$ consisting of those states $\Phi \in \mathcal{S}(M, \Delta, \leq_M)$ for which $\Phi^2(f) \in \mathcal{B}_f([\Phi]_1)$ for every face $f$ of $\Delta$ that is not a face of $\delta$.

Finally, for $\Phi \in \mathcal{S}(M, \Delta, \leq_M, \mathcal{B})$, we define

$$d(\Phi) = \left( \prod_{e \in Edge(\delta)} d(\Phi^1(e))^{\frac{1}{2}} \right) \left( \prod_{e \in Edge(\Delta) \setminus Edge(\delta)} d(\Phi^1(e)) \right)$$

and we have the following proposition.

PROPOSITION 2.2.3 *Let $M$ be a compact oriented 3-manifold with $\partial M = \Sigma$. For $\delta, \leq, \Delta, \leq_M, \mathcal{B}$ as above, and $\phi \in \mathcal{S}(\delta, \leq)$, define*

$$\begin{aligned} < (M, \Delta, \leq_M, \mathcal{B}); \phi > &= w^{-|Vert(\Delta)| + \frac{1}{2}|Vert(\delta)|} \\ &\times \sum_{\substack{\Phi \in \mathcal{S}(\Delta, \leq_M, \mathcal{B}) \\ \Phi|_\delta = \phi}} d(\Phi) \left( \prod_{T \in \Delta} \tilde{Z}(\Phi_T) \right) . \end{aligned} \qquad (2.2.11)$$

*Then,*

*(a) $< (M, \Delta, \leq_M, \mathcal{B}); \phi >$ depends only on $(M, \delta, \leq; \phi)$; i.e., this scalar does not depend on the choices of (i) the extension $\Delta$ of $\delta$, (ii) the extension $\leq_M$ of $\leq$, and (iii) the bases $\mathcal{B}$; so we may write $< (M, \delta, \leq); \phi >$ instead of $< (M, \Delta, \leq_M, \mathcal{B}); \phi >$;*

*(b) there exists a unique element $\xi_{(M,\delta,\leq)} \in W(\Sigma, \delta, \leq)$ such that*

$$< (M, \delta, \leq); \phi > \;=\; \langle \xi_{(M,\delta,\leq)}, \xi_\phi \rangle \;\; \forall \phi \in \mathcal{S}(\delta, \leq) \; ; \qquad (2.2.12)$$

*i.e., $\xi_\phi \mapsto < (M, \delta, \leq); \phi >$ extends uniquely to a conjugate-linear functional on $W(\Sigma, \delta, \leq)$.*

**Proof:** (a) Note that the case when $\Sigma$ is empty has already been established. The proof for non-trivial $\Sigma$ is essentially identical, the only difference being that we have to use a *relative version* - see[TV] - of Pachner's theorem, whose statement differs from that of Theorem 2.2.2 only in these two modifications:

(i) the triangulations $\Delta$ and $\Delta'$ are required to agree on $\partial M$, i.e., $\Delta \cap \partial M \;=\; \Delta' \cap \partial M$; and

(ii) the ambient isotopy is required to be *rel boundary* - i.e., one requires that the isotopy $f$ satisfies $f(x,t) = x \, \forall x \in \partial M, \ t \in [0,1]$.

(b) Since $\xi_\phi = \otimes_{f \in Face(\delta)} \tilde{\xi}_f$, where we write $\xi_f = \phi^2(f)$, and since the tensor product is 'multilinear in its factors', it is enough to verify the following: if $\phi, \psi \in \mathcal{S}(\Sigma, \delta)$ satisfy

$$\phi^1 = \psi^1, \phi^2(f) = \psi^2(f) \ \forall f \neq f_0, \phi^2(f_0) = \xi, \ \psi^2(f_0) = \eta$$

and if we define $\chi \in \mathcal{S}(\delta, \leq)$ by $\chi^1 = \phi^1, \chi^2(f) = \phi^2(f) \ \forall f \neq f_0$ and $\chi^2(f) = \xi + \alpha\eta$, where $\alpha \in \mathbb{C}$, then

$$< (M, \delta, \leq); \chi > = < (M, \delta, \leq); \phi > + \bar{\alpha} < (M, \delta, \leq); \psi > \ .$$

The only point in this statement is the following: if $T$ is a tetrahedron in $\Delta$ and one of the faces $f_0$ of $T$ lies in $\Sigma$, then $f_0$ is a negative face of $T$ (regarded as having the orientation coming from $M$); and this is easily verified. This will complete the proof of (b) since the hypotheses ensure the existence of a constant $C$ such that $< (M, \delta, \leq); \chi > = C\tilde{Z}(\chi_T)$ and $\tilde{Z}(\chi_T)$ depends conjugate linearly on the labels of negative faces. □

In the sequel, until we establish that $< (M, \delta, \leq); \chi >$ is independent of $\leq$, we shall find it convenient to adopt the notation of (2.2.10); thus, if $\leq$ and $\leq'$ are two total orders on $Vert(\delta)$ and if $\phi \in \mathcal{S}(\delta, \leq)$, we shall write $\phi'$ for the unique element of $\mathcal{S}(\delta, \leq')$ such that $U_{\leq, \leq'} \xi_\phi = \xi_{\phi'}$.

LEMMA 2.2.4 *Let $\Sigma, \delta, \leq, M, \Delta$ be as above, and suppose $\leq_M$ is another total order on $Vert(\delta)$. Then,*

$$< (M, \delta, \leq); \phi > \ = \ < (M, \delta, \leq'); \phi' > \ \forall \phi \in \mathcal{S}(\delta, \leq) \ .$$

**Proof:** Let $\leq_M$ (resp., $\leq'_M$) be any total order on $Vert(\Delta)$ which extends $\leq$ (resp., $\leq'$). The point is that if $\mathcal{B}$ is some choice of bases for $\Delta, \leq_M$, then there exists a corresponding choice $\mathcal{B}'$ of bases for $\Delta, \leq'_M$ and a bijection

$$\{\Phi \in \mathcal{S}(\Delta, \leq_M, \mathcal{B}) : \Phi|_\delta = \phi\} \leftrightarrow \{\Phi' \in \mathcal{S}(\Delta, \leq'_M, \mathcal{B}') : \Phi'|_\delta = \phi'\}$$

(defined using the Frobenius reciprocity maps as before) such that $d(\Phi) = d(\Phi')$ and $\tilde{Z}(\Phi_T) = \tilde{Z}(\Phi'_T)$ for all $\Phi \in \mathcal{S}(\Delta, \leq_M, \mathcal{B})$

such that $\Phi|_\delta = \phi$. (This is argued exactly as in the case of closed 3-manifolds $M$.)

□

We next wish to observe appropriate multiplicative and involutive properties of $\Sigma \mapsto W(\Sigma, \delta, \leq)$. For this, we shall find the following notation useful:

(i) If $\Sigma_i, \delta_i, \leq_i, i = 1, 2$ are as above, let us write $\delta_1 \coprod \delta_2$ for the obvious 'disjoint union triangulation' of the surface $\Sigma_1 \coprod \Sigma_2$. Let us also write $\leq_1 \coprod \leq_2$ for any total order on the vertex set of $\delta_1 \coprod \delta_2$ which restricts to $\leq_i$ on $Vert(\delta_i)$, for $i = 1, 2$. Finally if $\phi_i \in \mathcal{S}(\delta_i, \leq_i)$, we shall write $\phi_1 \coprod \phi_2$ for the obvious surface-state in $\mathcal{S}(\delta_1 \coprod \delta_2, \leq_1 \coprod \leq_2)$.

(ii) If $\Sigma, \delta, \leq$ are as above, then $\delta$ may also be regarded as a triangulation of $\Sigma^*$.

PROPOSITION 2.2.5 *(a) There exists a unique unitary isomorphism*

$$\begin{aligned} W(\Sigma_1 \coprod \Sigma_2, \delta_1 \coprod \delta_2, \leq_1 \coprod \leq_2) \\ \cong \; W(\Sigma_1, \delta_1, \leq_1) \otimes W(\Sigma_2, \delta_2, \leq_2) \end{aligned}$$

*which maps* $\xi_{\phi_1 \coprod \phi_2}$ *onto* $\xi_{\phi_1} \otimes \xi_{\phi_2}$ *for all* $\phi_i \in \mathcal{S}(\delta_i, \leq_i), i = 1, 2$.

*(b) There exists a unique unitary isomorphism*

$$W(\Sigma^*, \delta, \leq) \;\cong\; \overline{W(\Sigma, \delta, \leq)}$$

*which maps* $\xi_\phi$ *onto* $\overline{\xi_\phi}$ *for all* $\phi \in \mathcal{S}(\delta, \leq)$.

**Proof:** Both the statements follow from the definitions of the Hilbert spaces in question and Remark 1.2.5. □

PROPOSITION 2.2.6 *If* $(\Sigma_i, \delta_i, \leq_i)$, $i = 1, 2$ *are as above, and if* $M$ *is a cobordism from* $\Sigma_1$ *to* $\Sigma_2$, *then there exists a unique linear map* $Z_{(M,(\delta_1,\leq_1),(\delta_2,\leq_2))} : W(\Sigma_1, \delta_1, \leq_1) \to W(\Sigma_2, \delta_2, \leq_2)$ *such that*

$$\begin{aligned} &\langle Z_{(M,(\delta_1,\leq_1),(\delta_2,\leq_2))}(\xi_{\phi_1}), \xi_{\phi_2} \rangle \\ &\quad = \; < (M, \delta_1 \coprod \delta_2, \leq_1 \coprod \leq_2); \phi_1 \coprod \phi_2 > \;, \quad (2.2.13) \end{aligned}$$

*for all* $\phi_i \in \mathcal{S}(\delta_i, \leq_i), i = 1, 2$.

**Proof:** Begin by noting that $\partial M = \Sigma_1^* \coprod \Sigma_2$, and that the map

$$\begin{aligned}(\overline{W(\Sigma_1, \delta_1, \leq_1)} \otimes W(\Sigma_2, \delta_2, \leq_2)) &\ni \overline{\xi_{\phi_1}} \otimes \xi_{\phi_2} \\ \mapsto \quad & \xi_{\phi_1 \coprod \phi_2} \in W(\Sigma_1^* \coprod \Sigma_2, \delta_1 \coprod \delta_2, \leq_1 \coprod \leq_2) \\ \mapsto \quad & < (M, \delta_1 \coprod \delta_2, \leq_1 \coprod \leq_2); \phi_1 \coprod \phi_2 >\end{aligned}$$

is a conjugate-linear map, in view of Propositions 2.2.5(b) and 2.2.3(b); hence the passage from $(\xi_{\phi_1}, \xi_{\phi_2})$ to the right side of equation (2.2.13) defines a sesquilinear map $W(\Sigma_1, \delta_1, \leq_1) \times W(\Sigma_2, \delta_2, \leq_2) \to \mathbb{C}$; the desired conclusion follows immediately. □

COROLLARY 2.2.7 *Let $(\Sigma_i, \delta_i, \leq_i), i = 1, 2, 3$ be as above, and suppose $M$ (resp., $N$) is a cobordism from $\Sigma_1$ to $\Sigma_2$ (resp., from $\Sigma_2$ to $\Sigma_3$). Then,*

*(a)* $Z_{(M^*, (\delta_2^*, \leq_2), (\delta_1^*, \leq_1))} = Z^*_{(M, (\delta_1, \leq_1), (\delta_2, \leq_2))}$; *and*

*(b)* $Z_{(N \circ M, (\delta_1, \leq_1), (\delta_3, \leq_3))} = Z_{(N, (\delta_2, \leq_2), (\delta_3, \leq_3))} Z_{(M, (\delta_1, \leq_1), (\delta_2, \leq_2))}$.

**Proof:** (a) is an immediate consequence of Proposition 2.2.6.

(b) In the discussion so far, we have never encountered the need for the half-powers that appeared in equation (2.2.11); that makes its appearance now. Other than that, the verification of (b) is a straightforward - albeit probably somewhat tedious - consequence of the definitions.

□

We now proceed towards establishing the independence of the order $\leq$. This goes through the following lemma.

LEMMA 2.2.8 *Let $M, \Sigma_i, \delta_i, \leq_i, i = 1, 2$ be as in Lemma 2.2.6. Let $\leq_i'$ be another total order on $Vert(\delta_i)$ for $i = 1, 2$, and let $U^i_{\leq_i, \leq_i'} : W(\Sigma_i, \delta_i, \leq_i) \to W(\Sigma_i, \delta_i, \leq_i')$, $i = 1, 2$ be as in equation (2.2.9). Then the following is a commutative diagram:*

$$\begin{array}{ccc} W(\Sigma_1, \delta_1, \leq_1) & \stackrel{U^1_{\leq_1, \leq_1'}}{\rightarrow} & W(\Sigma_1, \delta_1, \leq_1') \\ Z_{(M, (\delta_1, \leq_1), (\delta_2, \leq_2))} \downarrow & & \downarrow Z_{(M, (\delta_1, \leq_1'), (\delta_2, \leq_2'))} \\ W(\Sigma_2, \delta_2, \leq_2) & \stackrel{U^2_{\leq_2, \leq_2'}}{\rightarrow} & W(\Sigma_2, \delta_2, \leq_2') \end{array}$$

**Proof:** Let $\phi \in \mathcal{S}(\delta_i, \leq_i)$, $i = 1,2$; then - using the 'priming convention' (whereby $U^i_{\leq_i,\leq'_i}\xi_{\phi_i} = \xi_{\phi'_i}$) - we see (by Proposition 2.2.6 and Lemma 2.2.4) that

$$\begin{aligned}
&\langle Z_{(M,(\delta_1,\leq'_1),(\delta_2,\leq'_2))}\xi_{\phi'_1}, \xi_{\phi'_2}\rangle \\
&= \; < (M, \delta_1 \coprod \delta_2, \leq'_1 \coprod \leq'_2); \phi'_1 \coprod \phi'_2 > \\
&= \; < (M, \delta_1 \coprod \delta_2, \leq_1 \coprod \leq_2); \phi_1 \coprod \phi_2 > \\
&= \; \langle Z_{(M,(\delta_1,\leq_1),(\delta_2,\leq_2))}\xi_{\phi_1}, \xi_{\phi_2}\rangle \,,
\end{aligned}$$

and the proof is complete. □

This lemma is our justification for dispensing henceforth with the total orders. Now we proceed towards dispensing with the triangulations; what is needed is the following lemma:

LEMMA 2.2.9 *Suppose $\delta_i, \delta'_i$ are two triangulations of a surface $\Sigma_i$, for $i = 1,2$. Consider the following operators:*

$$\begin{aligned}
p_{(\Sigma_i,\delta_i)} &= Z_{(\Sigma_i \times [0,1], \delta_i \coprod \delta_i)} \\
p_{(\Sigma_i,\delta'_i)} &= Z_{(\Sigma_i \times [0,1], \delta'_i \coprod \delta'_i)} \\
v_{\delta_i,\delta'_i} &= Z_{(\Sigma_i \times [0,1], \delta_i \coprod \delta'_i)} \,.
\end{aligned}$$

*Then,*

*(a) $p_{(\Sigma_i,\delta_i)}$ and $p_{(\Sigma_i,\delta'_i)}$ are projections;*
*(b) $v_{\delta_i,\delta'_i}{}^* v_{\delta_i,\delta'_i} = p_{(\Sigma_i,\delta_i)}$, and $v_{\delta_i,\delta'_i} v_{\delta_i,\delta'_i}{}^* = p_{(\Sigma_i,\delta'_i)}$;*
*(c) For any cobordism $M$ from $\Sigma_1$ to $\Sigma_2$, we have:*

$$Z_{(M,\delta_1,\delta_2)} p_{(\Sigma_1,\delta_1)} = p_{(\Sigma_2,\delta_2)} Z_{(M,\delta_1,\delta_2)} \,.$$

**Proof:** This lemma is a direct consequence of Corollary 2.2.7 (applied to the special cobordisms of the form $\Sigma \times [0,1]$).

□

Finally, if we define $V(\Sigma,\delta) = ran(p_{(\Sigma,\delta)})$, we see (from Lemma 2.2.9 (b)) that there exists a unitary operator $v_{(\delta,\delta')}$ : $V(\Sigma,\delta) \to V(\Sigma,\delta')$, for any two triangulations $\delta, \delta'$ of $\Sigma$, so that we may naturally associate a Hilbert space $V(\Sigma)$ to each surface; further, we see from Lemma 2.2.9 (c) that any cobordism $M \in Cob(\Sigma_1, \Sigma_2)$ induces a linear operator $Z_M : V(\Sigma_1) \to V(\Sigma_2)$. Thus we finally have our TQFT.

# Chapter 3

# von Neumann algebras

## 3.1 Preliminaries on $II_1$ factors

We begin by gathering some preliminary facts about $II_1$ factors, which are special kinds of von Neumann algebras. Recall that a von Neumann algebra is a $C^*$-algebra $M$ which admits a *pre-dual*: i.e., there exists some Banach space, say $X$, such that $M$ is, as a Banach space, isometrically isomorphic to the Banach dual space $X^*$. It is a fact that such a pre-dual is unique up to isometric isomorphism, and any choice of a pre-dual is usually denoted by $M_*$. It follows that $M$ has a canonical topology, usually called the *$\sigma$-weak* topology, which is the smallest vector space topology with respect to which evaluations at members of $M_*$ are continuous. It is customary to call a linear map between von Neumann algebras *normal* if it is continuous with respect to the $\sigma$-weak topologies on domain and range. The natural morphisms in the category of von Neumann algebras are the normal $*$-homomorphisms.

We record below some standard facts about von Neumann algebras.

(i) A $\sigma$-weakly closed $*$-closed subalgebra $N$ of a von Neumann algebra $M$ is also a von Neumann algebra, and $N$ is usually called a von Neumann subalgebra of $M$.

(ii) The algebra $\mathcal{L}(\mathcal{H})$, of all bounded operators on a Hilbert space, is a von Neumann algebra.

(iii) Any von-Neumann algebra is isomorphic (via a map which, together with its inverse, is a normal isomorphism) to a von Neumann subalgebra of $\mathcal{L}(\mathcal{H})$.

(iv) Any von Neumann algebra $M$ is the norm closed subspace spanned by the set $\mathcal{P}(M) = \{p \in M : p = p^* = p^2\}$ of its projections.

(v) If $p \in \mathcal{P}(Z(M))$ (where $Z(M) = \{x \in M : xy = yx \; \forall \; y \in M\}$ denotes the centre of $M$), then $Mp = \{xp : x \in M\}$ is a $\sigma$-weakly closed two-sided ideal in $M$; and more importantly, every $\sigma$-weakly closed two-sided ideal in $M$ is of this form (for a uniquely determined $p$).

(vi) Since $Z(M)$ is clearly a von Neumann subalgebra of $M$, it follows from (v) and (vi) (applied to $Z(M)$) that the following two conditions on $M$ are equivalent:

(a) $M$ has no non-trivial $\sigma$-weakly closed two-sided ideals; (i.e., $M$ is *topologically simple*;)

(b) $Z(M) = \mathbb{C}1$.

(vii) A von Neumann algebra is called a **factor** if it satisfies the two equivalent conditions of (vi). A factor is said to be a **finite factor** if it has no non-unitary isometry; i.e., if $u \in M$ and $u^*u = 1 \;\Rightarrow\; uu^* = 1$. It is a fact - see [JS], Proposition A.3.1, for instance - that a finite factor is even *algebraically* simple, meaning that it has no non-trivial two-sided ideals.

(viii) It is true that any finite-dimensional finite factor is isomorphic to $M_n(\mathbb{C})$ (equivalently, to $\mathcal{L}(\mathcal{H})$ for a finite-dimensional Hilbert space). In contrast, a $\mathbf{II_1}$ **factor** is just an infinite-dimensional finite factor.

(ix) Every finite factor admits a tracial state (i.e., a linear functional $\tau \in M^*$ such that $\tau(xy) = \tau(yx) \; \forall x, y \in M$, $\tau(1) = 1$ and $\tau(x^*x) \geq 0 \; \forall x \in M$). Further, such a $\tau$ is unique, and it is necessarily normal and faithful (i.e., $x \neq 0 \Rightarrow \tau(x^*x) > 0$). We shall simply write *tr* (or $tr_M$) to denote the unique trace on a finite factor.

We shall have occasion to use the following 'uniqueness' of not necessarily normalised traces on finite factors: if $\tau$ is a trace on $N$, then $\tau(\cdot) = \tau(1) \; tr_N(\cdot)$.

(x) For projections $p, q \in \mathcal{P}(M)$ ($M$ a $II_1$ factor), the following conditions are equivalent:

(a) there exists a unitary $u \in M$ (i.e., $uu^* = u^*u = 1$) such that $upu^* = q$;

(b) $tr\ p = tr\ q$.

Furthermore, $\{tr\ p : p \in \mathcal{P}(M)\} = [0,1]$.

(xi) The fundamental construction, as far as all representations of a finite factor $M$ are concerned, is that of the Hilbert space $L^2(M)$, which we outline. The equation $\langle x, y\rangle = tr(y^*x)$ defines a (positive-definite) inner product on $M$, and the completion of $M$, with respect to the norm coming from this inner product, is the Hilbert space $L^2(M)$. We shall write $\Omega$ for the vector in $L^2(M)$ which corresponds to the identity element 1 of $M$ (where $M$ is thought of as a dense subspace of $L^2(M)$). It can be shown that the natural $M - M$ **bimodule structure** of the algebra $M$ extends to an $M - M$ bimodule structure on the Hilbert space $L^2(M)$. (This just means that there exist (linear) normal *-preserving maps $\pi_l, \pi_r : M \to \mathcal{L}(L^2(M))$ such that $\pi_l$ preserves products, $\pi_r$ reverses products, and $\pi_l(x)\pi_r(y) = \pi_r(y)\pi_l(x)\ \forall x, y \in M$.) It should be clear that $\Omega$ is a cyclic vector for $\pi_l$ (meaning that $\{\pi_l(x)\Omega : x \in M\}$ is a dense subspace of $L^2(M)$), and that

$$\langle \pi_l(x)\Omega, \Omega\rangle = tr_M(x) .$$

(xii) Suppose $M$ is a $II_1$ factor with separable pre-dual (i.e., $M_*$ is separable, as a Banach space). (We shall only regard such von Neumann algebras in the rest of these notes.) For convenience, we shall identify $M$ as a von Neumann subalgebra of $\mathcal{L}(\mathcal{H}_1)$ (where we write $\mathcal{H}_1 = L^2(M)$) via the left-regular representation $\pi_l$ discussed in (xi) above. A bounded operator on a direct sum of infinitely many copies of $\mathcal{H}_1$ may naturally be identified with an infinite matrix whose entries are bounded operators on $\mathcal{H}_1$. We shall write $M_\infty(M)$ for the set of those infinite matrices $((x_{ij}))_{1\le i,j<\infty}$ which define bounded operators on a direct sum of infinitely many copies of $\mathcal{H}_1$, for which $x_{ij} \in M\ \forall i, j$. We shall also write $\mathcal{H}_\infty$ for the direct sum of countably infinitely many copies of $\mathcal{H}_1$, which we shall think of as the set of infinite row vectors $\xi = (\xi_1, \xi_2, \cdots, \xi_n, \cdots)$ with $\sum_n ||\xi_n||^2 < \infty$. Then it is easily seen that $\mathcal{H}_\infty$ is naturally an $M - M_\infty(M)$ bimodule via matrix-multiplication (and the $M - M$ bimodule structure on $L^2(M)$). For any projection $q = ((q_{ij})) \in \mathcal{P}(M_\infty(M))$, we shall write

$$Tr\ q = \sum_{i=1}^{\infty} tr_M(q_{ii}) \ .$$

(Note that $q_{ii} \geq 0$ and so $tr(q_{ii}) \geq 0$, and hence $Tr\ q \in [0, \infty]$.)

Observe now that $\mathcal{H}_q = \mathcal{H}_\infty \cdot q$ is naturally a left $M$-module. The following two facts are fundamental to our further study of (bi)modules.

(a) If $q_1, q_2 \in \mathcal{P}(M_\infty(M))$, then

$$_M\mathcal{H}_{q_1} \cong {_M\mathcal{H}_{q_2}} \Leftrightarrow Tr\ q_1 = Tr\ q_2 \ .$$

(b) Every separable $M$-module is isomorphic to some $\mathcal{H}_q$. Above and elsewhere, two $M$-modules are called isomorphic if there exists a unitary $M$-linear operator of one onto the other. Further, if an $M$-module $X$ is as in (b) above, then the so-called **M dimension** of $X$ is denoted by $dim\ _MX$, and is defined by $dim\ _MX = Tr\ q$.

(xiii) With $\mathcal{H}_q$ as above, every operator in $End\ (_M\mathcal{H}_q)$ is of the form $\xi \mapsto \xi \cdot b$ for a unique $b \in qM_\infty(M)q$ (i.e., for a unique $b \in M_\infty(M)$ which satisfies $b = qbq$).

(xiv) A vector $\xi$ in an $M$-module $X$ is said to be a **bounded vector** if the following equivalent conditions are satisfied:

(a) there exists a constant $K > 0$ such that

$$||x \cdot \xi||^2 \leq K\ tr_M(x^*x)\ \forall\ x \in M \ ;$$

(b) there exists a (unique) $M$-linear operator $R_\xi : {_ML^2(M)} \to {_MX}$ such that $R_\xi(\Omega) = \xi$.

(xv) The set of bounded vectors in an $M$-module $X$ is denoted by the symbol $X^0$. Then, we have:

(a) the subspace $X^0$ is $M$-stable and dense; further

$$T(X^0) \subseteq Y^0\ \forall\ T \in Hom(_MX, {_MY}) \ ;$$

(b) conversely, any '$M$-linear' map from $X^0$ to $Y^0$ extends to a unique $M$-linear bounded operator in $Hom(_MX, {_MY})$.

# 3.2 Bases and $M$-valued inner products

Henceforth symbols such as $M$, $N$, $P$, $Q$ will always denote $II_1$ factors with separable preduals, and symbols such as $X$, $Y$, $Z$, $A$, $B$, $C$, $D$ will always denote left- or/and right- modules. For instance, if we write ${}_PX_Q$, then this represents a $P-Q$ bimodule $X$. Also, all our modules will be of **finite type** - meaning that $dim_P\ {}_PX, dim_Q\ X_Q < \infty$.

Assume, until further notice, that we have fixed an ${}_MX$. In view of fact (xii)(a) of the previous section and the assumption that $dim_M X < \infty$, we may (choose some integer $n$ which is at least as large as $dim_M X$, and) assume that there exists a projection $q \in \mathcal{P}(M_n(M))$ such that $X = M_{1\times n}(L^2(M)) \cdot q$. Thus, a typical member of $X$ is of the form $\xi = [\xi_1, \cdots, \xi_n] \in M_{1\times n}(L^2(M))$ which satisfies $\xi = \xi \cdot q$ - meaning, of course, that $\xi_j = \sum_{i=1}^n \xi_i \cdot q_{ij}$. It is a fact that, in this case, $\xi \in X^0$ if and only if $\xi_j \in M\ \forall j$ (where we regard $M$ as a dense subspace of $L^2(M)$). Thus

$$X^0 = M_{1\times n}(M) \cdot q \subset M_{1\times n}(L^2(M)) \cdot q \ = X\ .$$

Let us write $q^i = [q_{i1}, \cdots, q_{in}]$ for the $i$-th row of $q$. The idempotence of $q$ implies that $q^i \in M_{1\times n}(M) \cdot q$; thus $q^i \in X^0\ \forall i$. Now, for any $\xi = [\xi_1, \cdots, \xi_n] \in X^0$, and $1 \leq j \leq n$, note that

$$\begin{aligned} \xi_j &= \sum_{i=1}^n \xi_i \cdot q_{ij} \\ &= \sum_{i=1}^n \pi_r(q_{ij})(\xi_i) \\ &= \sum_{i=1}^n \xi_i q_{ij} \\ &= \sum_{i=1}^n \pi_l(\xi_i) q_{ij} \\ &= \sum_{i=1}^n (\xi_i \cdot q^i)_j\ , \end{aligned}$$

and hence,

$$\xi = \sum_{i=1}^{n} \xi_i \cdot q^i \ \forall \ \xi \in X^0 \ . \tag{3.2.1}$$

PROPOSITION 3.2.1 *(a) For any $X = {}_M X$, there exists a unique mapping - called the $M$-***valued inner product on** $X$ *- denoted by*

$$X^0 \times X^0 \ \ni \ (\xi, \eta) \ \mapsto \ {}_M\langle \xi, \eta \rangle \ \in M$$

*which satisfies, for all $\xi$, $\xi_1, \xi_2, \eta \in X^0$, $m \in M$:*

*(i)* ${}_M\langle \xi, \xi \rangle \ \geq \ 0;$

*(ii)* ${}_M\langle \xi, \eta \rangle \ = \ {}_M\langle \eta, \xi \rangle^*;$

*(iii)* ${}_M\langle m \cdot \xi_1 + \xi_2, \eta \rangle \ = \ m \ {}_M\langle \xi_1, \eta \rangle \ + \ {}_M\langle \xi_2, \eta \rangle;$

*(iv)* $\langle \xi, \eta \rangle \ = \ tr_M( \ {}_M\langle \xi, \eta \rangle \ ).$

*(b) If $T \in Hom({}_M X, {}_M Y)$, then*

$${}_M\langle T(\xi), \eta \rangle \ = \ {}_M\langle \xi, T^*(\eta) \rangle \ , \forall \ \xi \in X^0, \ \eta \in Y^0 \ .$$

*In particular, if it so happens that $X = {}_M X_N$ is a bimodule (of finite type), then it follows automatically that*

$${}_M\langle \xi \cdot n, \eta \rangle \ = \ {}_M\langle \xi, \eta \cdot n^* \rangle \ , \forall \ \xi, \eta \in X^0 \text{ and } n \in N \ .$$

**Proof**: (a) We may assume, without loss of generality, that $X = \mathcal{H}_q$ as above, for some $q \in \mathcal{P}(M_n(M))$.

For existence, if $\xi, \eta \in X^0 = M_{1 \times n}(M) \ q$, we may define

$${}_M\langle \xi, \eta \rangle \ = \ \xi \eta^* \ = \ \sum_{i=1}^{n} \xi_i \eta_i^* \ , \tag{3.2.2}$$

and easily verify that the four conditions above are satisfied.

For uniqueness, suppose ${}_M\langle\langle \cdot, \cdot \rangle\rangle$ is another such $M$-valued inner product defined on $X^0 \times X^0$, with $X$ as above. Notice first that, if $q^i$ is the $i$-th row of $q$ as above, then

$${}_M\langle q^i, q^j \rangle = \sum_k q_{ik} q_{jk}^* = q_{ij} \ , \tag{3.2.3}$$

since $q = q^* = q^2$. Define $p_{ij} = {}_M\langle\langle q^i, q^j\rangle\rangle$; then $p = ((p_{ij})) \in M_n(M)$. Since $(q^m)_r = q_{mr}$, it then follows from equation (3.2.1) that

$$\begin{aligned} p_{ij} &= {}_M\langle\langle q^i, q^j\rangle\rangle \\ &= {}_M\langle\langle \sum_{k=1}^n q_{ik} \cdot q^k, \sum_{l=1}^n q_{jl} \cdot q^l \rangle\rangle \\ &= \sum_{k,l=1}^n q_{ik}\ {}_M\langle\langle q^k, q^l\rangle\rangle q_{jl}^* \quad (*) \\ &= \sum_{k,l=1}^n q_{ik} p_{kl} q_{lj} \ , \end{aligned}$$

where, in the step marked (*), we have used the following easy consequence of the conditions (ii) and (iii) to be satisfied by the $M$-valued inner product ${}_M\langle\langle\cdot,\cdot\rangle\rangle$:

$${}_M\langle\langle \xi, a \cdot \eta\rangle\rangle = {}_M\langle\langle \xi, \eta\rangle\rangle\ a^* \ .$$

Hence, $p = qpq \in qM_n(M)q$. It follows from (xiii) of the last subsection that there exists a unique operator $P \in End({}_M X)$ such that $P\xi = \xi p$ (matrix-multiplication) for each $\xi \in X$.

Notice then that, for any $a, b \in M$, we have

$$\begin{aligned} \langle P(a \cdot q^i), (b \cdot q^j)\rangle &= tr_M\ ({}_M\langle a \cdot Pq^i, b \cdot q^j\rangle\ ) \\ &= tr_M\ (a\ (q^i p)(q^j)^*\ b^*) \\ &= tr_M(a(qpq)_{ij} b^*) \\ &= tr_M(a p_{ij} b^*) \\ &= tr_M(a\ {}_M\langle\langle q^i, q^j\rangle\rangle b^*) \\ &= \langle a \cdot q^i, b \cdot q^j\rangle \end{aligned}$$

and consequently, by equation (3.2.1),

$$\langle P\xi, \eta\rangle = \langle \xi, \eta\rangle\ , \forall \xi, \eta \in X^0 \ .$$

Since $X^0$ is dense in $X$, it follows now that $P = id_X$.

It follows from the uniqueness assertion of (xiii) of the last subsection, and the fact that $q$ is the identity of $qM_n(M)q$, that

$p = q$; i.e., ${}_M\langle\langle q^i, q^j\rangle\rangle = {}_M\langle q^i, q^j\rangle$ for all $i, j$. This and equation (3.2.1) now show that ${}_M\langle\langle \xi, \eta\rangle\rangle = {}_M\langle \xi, \eta\rangle$ for all $\xi, \eta \in X^0$.

(b) Firstly, suppose $T \in \mathcal{L}(X, Y)$ and has (left) polar decomposition $T = U|T|$. It is then true that $T \in Hom({}_MX, {}_MY)$ if and only if $U \in Hom({}_MX, {}_MY)$ and $|T| \in End({}_MX)$. We omit the proof of this fact since it is exactly similar to the proof of the fact that a von Neumann algebra contains an operator if and only if it contains the 'parts' of the polar decomposition. (This latter fact contains the fact displayed above, in case $X = Y$.)

With $U$ as above, where $T \in Hom({}_MX, {}_MY)$, it should be clear that $ran\, U$ and $ran\, U^*$ are $M$-modules, and that $U$ defines an $M$-linear unitary operator between these spaces. Further, since every $C^*$-algebra (and, in particular, $End({}_MX)$) is linearly spanned by its unitary elements, it suffices to prove the assertion in (b) in the special case when $T$ is unitary. (*Reason:* If this has been established, the truth of (b) will have been established with $T$ replaced by $U$ or $|T|$, and we could conclude that

$$\begin{aligned} {}_M\langle T\xi, \eta\rangle &= {}_M\langle U|T|\xi, \eta\rangle \\ &= {}_M\langle |T|\xi, U^*\eta\rangle \\ &= {}_M\langle \xi, |T|U^*\eta\rangle \\ &= {}_M\langle \xi, T^*\eta\rangle \ . ) \end{aligned}$$

In case $T$ is unitary, the equation

$${}_M\langle\langle \xi, \eta\rangle\rangle = {}_M\langle T\xi, T\eta\rangle$$

is easily verified to define an assignment ${}_M\langle\langle \cdot, \cdot\rangle\rangle$ which satisfies the requirements (i)-(iii) of an $M$-valued inner product. As for (iv), the assumed unitarity of $T$ implies that if $\xi, \eta \in X^0$, then

$$\begin{aligned} tr_M(\ {}_M\langle T\xi, T\eta\rangle\ ) &= \langle T\xi, T\eta\rangle \\ &= \langle \xi, \eta\rangle \end{aligned}$$

and hence the uniqueness of the $M$-valued inner product now implies that ${}_M\langle T\xi, T\eta\rangle = {}_M\langle \xi, \eta\rangle$, from which the desired equality follows easily.

□

In the language of the $M$-valued inner product, we may rewrite equation (3.2.1) as

$$\xi = \sum_{i=1}^{n} {}_M\langle \xi, q^i \rangle \cdot q^i \ , \ \forall \xi \in X^0 \ . \tag{3.2.4}$$

Any finite collection $\{q^i\}_i \subset X^0$ which satisfies equation (3.2.4) will be called a **left-basis** for ${}_MX$. It is not hard to show, conversely, that if $\{\eta_i : 1 \leq i \leq m\}$ is a left-basis for an $M$-module ${}_MY$, and if we define $q_{ij} = {}_M\langle \eta_i, \eta_j \rangle$, then $q = ((q_{ij})) \in \mathcal{P}(M_m(M))$ and that ${}_MY$ is isomorphic to $M_{1\times m}(L^2(M)) \cdot q$. If $\{\eta_j\}$ is such a left-basis for ${}_MY$, then

$${}_M\langle \xi, \eta \rangle = \sum_j {}_M\langle \xi, \eta_j \rangle \ {}_M\langle \eta_j, \eta \rangle \ , \ \forall \ \xi, \eta \in Y^0 \ . \tag{3.2.5}$$

Notice also that for such $\{\eta_i\}_i$, we have:

$$\sum_i ||\eta_i||^2 = \sum_i tr_M( \ {}_M\langle \eta_i, \eta_i \rangle \ ) = \sum_i Tr \ q_{ii} = dim \ {}_MX \ . \tag{3.2.6}$$

## 3.3 Contragredients and tensor products

Recall, to start with, that an **anti-unitary** operator between Hilbert spaces is a norm-preserving conjugate-linear surjection. If $J : X \to Y$ is an anti-unitary operator, it is customary to write $J^*$ for the inverse mapping $J^{-1}$, and it is easily verified that

$$\langle J^*\eta, \xi \rangle \ = \ \langle J\xi, \eta \rangle$$

for all $\xi \in X, \eta \in Y$.

DEFINITION 3.3.1 *If ${}_MX$ is a left $M$-module, then a right $M$-module $\bar{X}_M$ is called (a model of) the* **contragredient** *module of ${}_MX$ if there exists an anti-unitary operator $X \ni \xi \mapsto \bar{\xi} \in \bar{X}$ such that*

$$\overline{m \cdot \xi} \ = \ \bar{\xi} \cdot m^* \ \forall \ \xi \in X, m \in M \ .$$

REMARK 3.3.2 (a) It should be obvious that (a model of) the contragredient module exists (simply define the right $M$-module structure on the conjugate Hilbert space $\bar{X}$ by the equation in the above definition), and that any two models are isomorphic (since the composition of two anti-unitary operators is a unitary operator). So we shall henceforth dispense with the awkward phrase 'a model of' and simply talk of *the* contragredient module.

(b) With the notation of the above definition, we shall also refer to ${}_MX$ as the contragredient of $\bar{X}_M$. Similarly, we shall say that a bimodule ${}_N\bar{X}_M$ is the contragredient of ${}_MX_N$ if there exists an anti-unitary operator $X \ni \xi \mapsto \bar{\xi} \in \bar{X}$ satisfying

$$\overline{m \cdot \xi \cdot n} = n^* \cdot \bar{\xi} \cdot m^* \ \forall \ \xi \in X, m \in M \ , n \in N.$$

(c) If $X_M$ is a right $M$-module, then ${}_M\bar{X}$ is a left $M$-module, and it is easy to see that

$$\xi \in (X_M)^0 \ \Leftrightarrow \ \bar{\xi} \in ({}_M\bar{X})^0 \ .$$

Further, if we define

$$\langle \xi, \eta \rangle_M \ = \ {}_M\langle \bar{\xi}, \bar{\eta} \rangle \ \forall \ \xi, \eta \in X^0 \ , \tag{3.3.7}$$

then it is easily verified that this *right-handed $M$-valued inner product* (which is conjugate-linear in the first variable and linear in the second variable) is the unique such mapping, which satisfies the following conditions, for all $\xi, \xi_1, \xi_2, \eta \in X^0$, $m \in M$:

(i) $\langle \xi, \xi \rangle_M \ \geq \ 0$;

(ii) $\langle \xi, \eta \rangle_M \ = \ (\langle \eta, \xi \rangle_M)^*$;

(iii) $\langle \xi, \eta_1 \cdot m + \eta_2 \rangle_M \ = \ \langle \xi, \eta_1 \rangle_M m \ + \ \langle \xi, \eta_2 \rangle_M$;

(iv) $\langle \xi, \eta \rangle \ = \ tr_M( \ \langle \eta, \xi \rangle_M \ )$.

A finite collection $\{\xi_i\} \subset (X_M)^0$ in a right-module is called a right-basis for $X_M$ if $\xi = \sum_i \xi_i \cdot \langle \xi_i, \xi \rangle_M \ \forall \ \xi \in (X_M)^0$. As in equation (3.2.6), we have $\sum_i ||\xi_i||^2 = dim \ X_M$ for any right-basis of $X_M$.

We now pass to the tensor-product construction. We begin with some observations on bimodules. Suppose, then, that ${}_MX_N$ is an $M - N$ bimodule of finite type. Then, we may assume that $X \ = \ M_{1 \times m}(L^2(M)) \cdot p$ for some $m$ and $p$, with the natural (diagonal) left action of $M$. Since the right $N$-action commutes with the left $M$-action, it follows from §3.1 (xiii) that there exists a map $\theta : N \to pM_m(M)p$ such that $\xi \cdot b = \xi\theta(b)$ for all $\xi \in X$ and $b \in N$. It is easily verified that $\theta$ is a *-homomorphism which maps the 1 of $N$ to $p$. We shall use the symbol $X = \mathcal{H}_{p,\theta}$ to indicate that $X$ is a bimodule of finite type which has this description. In this case, we shall write $\theta(b) = ((\theta_{ij}(b)))$. The fact that $\theta$ has the properties stated a few lines above translates as follows:

(i) $\theta_{ij}(b^*) = \theta_{ji}(b)^*$;
(ii) $\theta_{ij}(bb') = \sum_s \theta_{is}(b)\theta_{sj}(b')$;
(iii) $\theta_{ij}(1) = p_{ij}$.

REMARK 3.3.3 For a bimodule ${}_MX_N$ of finite type, we have $({}_MX)^0 = (X_N)^0$; thus a vector in a bimodule of finite type is a bounded vector for the left action precisely when it is a bounded vector for the right action, and so the notion of 'bounded vector' makes unambiguous sense for bimodules of finite type.

To see this, we may assume that $X = \mathcal{H}_{p,\theta}$ as above. By symmetry, it suffices to check that $({}_MX)^0 \subset (X_N)^0$. (*Reason:* Any 'right-statement' for $X_M$ is equivalent to a corresponding 'left-statement' for ${}_M\bar{X}$.) In fact, it even suffices to verify that $p^i \in (X_N)^0\ \forall i$. (Reason: If this has been verified, and if $a \in M$, it follows from §3.1(xv)(a) - applied to $X_N$ and the operator of left-multiplication by $a$ - that also $a \cdot p^i \in (X_N)^0$; since $\xi = \sum_i \xi_i \cdot p^i$, this will do the trick.) So fix $p^i$ and $b \in N$, note that $p^i\theta(b)$ is the $i$-th row of $p\theta(b) = \theta(b)$, and hence

$$\begin{aligned}
||p^i \cdot b||^2 &= ||p^i\theta(b)||^2 \\
&= ||\ [\theta_{i1}(b), \theta_{i2}(b), \cdots, \theta_{im}(b)]\ ||^2 \\
&= \sum_{j=1}^{m} tr_M(\theta_{ij}(b)\theta_{ij}(b)^*) \\
&= tr_M(\ \sum_j \theta_{ij}(b)\theta_{ji}(b^*)) \\
&= tr_M(\ \theta_{ii}(bb^*)) \\
&\leq Tr\ \theta(bb^*) \\
&= (Tr\ \theta(1))tr_N(bb^*)\ ,
\end{aligned}$$

where we have used the 'uniqueness' of traces on finite factors - see §3.1 (ix).

Suppose now that, in addition to ${}_MX_N$, we are also given an $N$ module ${}_NY$ of finite type. We may assume, then, that $X = \mathcal{H}_{p,\theta}$ - with $\theta(1_N) = p \in \mathcal{P}(M_m(M))$ - and that $Y = M_{1\times n}(L^2(N)) \cdot q$. Let $\{p^i : 1 \leq i \leq m\}$ and $\{q^k : 1 \leq k \leq n\}$ be the associated left-bases of $X$ and $Y$ respectively. With this notation, define $r \in M_{mn}(M)$ by

$$r_{ik,jl} = \theta_{ij}(q_{kl}) .$$

It is fairly painless to verify that $r \in \mathcal{P}(M_{mn}(M))$. (For instance, the idempotence of $r$ follows from property (ii) of $\theta$ listed above, while the self-adjointness of $r$ is a consequence of property (i) above, thus:

$$(r_{ik,jl})^* = (\theta_{ij}(q_{kl}))^* = \theta_{ji}(q_{kl}^*) = \theta_{ji}(q_{lk}) = r_{jl,ik} \ .)$$

Notice that

$$\begin{aligned} Tr\ (r) &= \sum_{i=1}^{m}\sum_{k=1}^{n} tr_M(r_{ik,ik}) \\ &= \sum_{i=1}^{m}\sum_{k=1}^{n} tr_M(\theta_{ii}(q_{kk})) \\ &= \sum_{k=1}^{n} Tr(\theta(q_{kk})) \\ &= \sum_{k=1}^{n} Tr(\theta(1))\ tr_N(q_{kk}) \\ &= (Tr\ p) \sum_{k=1}^{n} tr_N(q_{kk}) \\ &= (Tr\ p)(Tr\ q) \\ &= (dim_M\ {}_MX)(dim_N\ {}_NY) . \end{aligned}$$

Thus, we see that $M_{1\times mn}(L^2(M)) \cdot r$ is a left $M$-module, say ${}_MZ$, with $dim_M Z = (dim_M X)(dim_N Y)$. This $Z$ is a model for the module $X \otimes_N Y$ occurring in the next proposition.

**Proposition 3.3.4** *Given a bimodule ${}_M X_N$ and a module ${}_N Y$ of finite type, there exists a module ${}_M(X \otimes_N Y)$ with the following properties:*

*(a) $(X \otimes_N Y)^0$ is the quotient of the algebraic tensor product $X^0 \otimes Y^0$ by the vector subspace spanned by $\{(\xi \cdot a) \otimes \eta - \xi \otimes (a \cdot \eta) : \xi \in X^0, a \in N, \eta \in Y^0\}$; we write $\xi \otimes_N \eta$ for the image of $\xi \otimes \eta$ in the above quotient; then it is the case that $a \cdot (\xi \otimes_N \eta) = (a \cdot \xi) \otimes_N \eta \ \forall \ a \in M$.*

*(b) $dim_M \ X \otimes_N Y = dim_M \ X \cdot dim_N \ Y$.*

*(c) If $\{\xi_i : 1 \leq i \leq m\}$ and $\{\eta_j : 1 \leq j \leq n\}$ are left-bases for $X$ and $Y$ respectively, then $\{\xi_i \otimes_N \eta_j : 1 \leq i \leq m, 1 \leq j \leq n\}$ is a left-basis for $X \otimes_N Y$.*

*(d) The (left) $M$-valued inner product on $X \otimes_N Y$ satisfies*

$$ {}_M\langle \xi \otimes_N \eta, \xi' \otimes_N \eta' \rangle = {}_M\langle \xi \cdot {}_N\langle \eta, \eta' \rangle, \xi' \rangle , \ \forall \, \xi, \xi' \in X^0, \ \eta, \eta' \in Y^0 . \tag{3.3.8} $$

**Proof:** With $X$ and $Y$ as in the preceding discussion, let $X \otimes_N Y$ be the $M$-module we called $Z$ in the line before the proposition. The validity of (b) has already been remarked.

For $\xi = [\xi_1, \cdots, \xi_m] \in M_{1 \times m}(M)p = X^0$, $\eta = [\eta_1, \cdots, \eta_n] \in M_{1 \times n}(N)q = Y^0$, let $\xi \otimes_N \eta = ((\ (\xi \otimes_N \eta)_{jl})) \in M_{1 \times mn}(M)r$ be defined by

$$ (\xi \otimes_N \eta)_{jl} = (\xi \cdot \eta_l)_j = (\xi \theta(\eta_l))_j = \sum_{i=1}^{m} \xi_i \theta_{ij}(\eta_l). $$

It is easily verified that this $\xi \otimes_N \eta$ indeed belongs to $M_{1 \times mn}(M)r$, and that $(\xi \cdot b) \otimes_N \eta = \xi \otimes_N (b \cdot \eta)$ for all $b \in N$. (For instance, $((\xi \cdot b) \otimes_N \eta)_{jl} = ((\xi \cdot b) \cdot \eta_l)_j = (\xi \cdot (b \cdot \eta)_l)_j = (\xi \otimes_N (b \cdot \eta))_{jl}$.) It is also easily seen that $(a \cdot \xi) \otimes_N \eta = a \cdot (\xi \otimes_N \eta) \ \forall a \in M$. It follows that the assignment $\xi \otimes \eta \mapsto \xi \otimes_N \eta$ defines a map of the *balanced tensor product* $X^0 \otimes_N Y^0$ (i.e., the quotient described in (a)) onto an $M$-stable linear subspace of $(X \otimes_N Y)^0$.

Another straightforward, although slightly more tedious, verification shows that $(p^i \otimes_N q^k) = r^{ik}$. This establishes (c) as well as the fact that the map $X^0 \otimes_N Y^0 \to (X \otimes_N Y)^0$, as defined in the last paragraph, is surjective.

As for (d), first consider the case when $\xi = p^i, \xi' = p^j, \eta = q^k, \eta' = q^l$; and note that

$$\begin{aligned} {}_M\langle p^i \otimes_N q^k, p^j \otimes_N q^l \rangle &= {}_M\langle r^{ik}, r^{jl} \rangle \\ &= r_{ik,jl} \\ &= (p^i \otimes_N q^k)_{jl} \\ &= (p^i \cdot (q^k)_l)_j \\ &= {}_M\langle p^i \cdot q_{kl}, p^j \rangle \\ &= {}_M\langle p^i \cdot {}_N\langle q^k, q^l \rangle, p^j \rangle \ , \end{aligned}$$

as desired. The extension from the case of basis vectors to general vectors is an easy exercise.

To complete the proof of the proposition, we only need to establish the injectivity of the map from the balanced tensor product $X^0 \otimes_N Y^0$ onto $(X \otimes_N Y)^0$; thus, we should show that if $\xi(1), \cdots, \xi(s) \in X^0, \eta(1), \cdots, \eta(s) \in Y^0$ and if $\sum_{t=1}^s \xi(t) \otimes_N \eta(t) = 0$, then $\sum_{t=1}^s \xi(t) \otimes \eta(t)$ is expressible as a sum of differences of the sort mentioned in the statement of (a). For this, note that

$$\begin{aligned} \sum_{t=1}^s \xi(t) \otimes \eta(t) &= \sum_{t=1}^s \xi(t) \otimes \left( \sum_{l=1}^n {}_N\langle \eta(t), q^l \rangle \cdot q^l \right) \\ &= \zeta + \sum_{l=1}^n \xi_l \otimes q^l \ , \end{aligned}$$

where

$$\zeta = \sum_{t=1}^s \sum_{l=1}^n [\xi(t) \otimes ({}_N\langle \eta(t), q^l \rangle \cdot q^l) - (\xi(t) \cdot {}_N\langle \eta(t), q^l \rangle) \otimes q^l]$$

and

$$\xi_l = \sum_{t=1}^s \xi(t) \cdot {}_N\langle \eta(t), q^l \rangle \ .$$

Since $\zeta$ is clearly a sum of the desired sort, we will be done as soon as we show that

$$\sum_{t=1}^s \xi(t) \otimes_N \eta(t) = 0 \ \Rightarrow \ \xi_l = 0 \ \forall \ l.$$

This is true because

$$\begin{aligned}
\xi_l &= \sum_{t=1}^{s} \xi(t) \cdot {}_N\langle \eta(t), q^l \rangle \\
&= \sum_{t=1}^{s} \xi(t) \cdot (\eta(t))_l \\
&= \sum_{t=1}^{s} \sum_{j=1}^{m} \left(\xi(t) \cdot (\eta(t))_l\right)_j \cdot p^j \\
&= \sum_{t=1}^{s} \sum_{j=1}^{m} \left(\xi(t) \otimes_N \eta(t)\right)_{jl} \cdot p^j \\
&= 0 \,,
\end{aligned}$$

and the proof of the proposition is complete. □

REMARK 3.3.5 (a) As is to be expected, the tensor product we have defined has a description in terms of the following universal property, whose proof follows quite easily from Proposition 3.3.4 and §3.1(xv):

Suppose ${}_M X_N, {}_N Y$ are as above, and that ${}_M Z$ is an $M$-module of finite type; assume further that there exists a bilinear map

$$X^0 \times Y^0 \ni (\xi, \eta) \mapsto B(\xi, \eta) \in Z^0$$

which is balanced, i.e., satisfies

$$B(a \cdot \xi \cdot b, \eta) = a \cdot B(\xi, b \cdot \eta) \; \forall a \in M, b \in N \; .$$

Then there exists a unique bounded operator $B_0 \in Hom({}_M(X \otimes_N Y), Z)$ such that $B(\xi, \eta) = B_0(\xi \otimes_N \eta)$.

This universal characterisation also shows that the tensor product is unique up to isomorphism as an $M$-module.

(b) If $S \in Hom({}_M X_N, {}_M X'_N)$ and $T \in Hom({}_N Y, {}_N Y')$, then there exists a unique $S \otimes_N T \in Hom({}_M(X \otimes_N Y), {}_M(X' \otimes_N Y'))$ (by (a) above) such that $(S \otimes_N T)(\xi \otimes_N \eta) = (S\xi \otimes_N T\eta)$ for all $\xi \in X^0, \eta \in Y^0$.

(c) It is true (and easily verified) that various natural relations hold, such as:

(i) $S \otimes_N T = (id_{X'} \otimes_N T) \circ (S \otimes_N id_Y) = (S \otimes_N id_{Y'}) \circ (id_X \otimes_N T)$;

(ii) $(S \otimes_N T)^* = S^* \otimes_N T^*$.

(d) If it turns out that $Y = {}_N Y_P$ is an $N - P$ bimodule of finite type, then $X \otimes_N Y$ is an $M - P$ bimodule of finite type, with the actions satisfying $a \cdot (\xi \otimes_N \eta) \cdot c = (a \cdot \xi) \otimes_N (\eta \cdot c)$ for all $a \in M, c \in P$. Also, it is easily verified that if $S$ and $T$ are $M - N$-linear and $N - P$-linear, respectively, then $S \otimes_N T$ is $M - P$-linear.

Furthermore, almost identical arguments to the ones given for the preceding proposition show that

(i) if $\{\xi_i : 1 \leq i \leq m\}$ and $\{\eta_j : 1 \leq j \leq n\}$ are right-bases for $X$ and $Y$ respectively, then $\{\xi_i \otimes_N \eta_j : 1 \leq i \leq m, 1 \leq j \leq n\}$ is a right-basis for $X \otimes_N Y$; and

(ii) we have

$$\langle \xi \otimes_N \eta, \xi' \otimes_N \eta' \rangle_P = \langle \eta, \langle \xi, \xi' \rangle_N \cdot \eta' \rangle_P . \tag{3.3.9}$$

PROPOSITION 3.3.6 *If ${}_P X_Q, {}_Q Y_R$ are bimodules of finite type, then the association $\overline{\xi \otimes_Q \eta} \mapsto \bar{\eta} \otimes_Q \bar{\xi}$ extends to an isomorphism $\overline{X \otimes_Q Y} \cong \bar{Y} \otimes_Q \bar{X}$ of $R - P$ bimodules.*

**Proof:** It clearly suffices to check that

$$_R\langle \overline{\xi \otimes_Q \eta}, \overline{\xi' \otimes_Q \eta'} \rangle = {}_R\langle \bar{\eta} \otimes_Q \bar{\xi}, \bar{\eta}' \otimes_Q \bar{\xi}' \rangle .$$

This is true because

$$\begin{aligned} _R\langle \overline{\xi \otimes_Q \eta}, \overline{\xi' \otimes_Q \eta'} \rangle &= \langle \xi \otimes_Q \eta, \xi' \otimes_Q \eta' \rangle_R \\ &= \langle \eta, \langle \xi, \xi' \rangle_Q \cdot \eta' \rangle_R \\ &= \langle \langle \xi', \xi \rangle_Q \cdot \eta, \eta' \rangle_R \\ &= \langle {}_Q\langle \bar{\xi}', \bar{\xi} \rangle \cdot \eta, \eta' \rangle_R \\ &= {}_R\langle \overline{{}_Q\langle \bar{\xi}', \bar{\xi} \rangle \cdot \eta}, \bar{\eta}' \rangle \\ &= {}_R\langle \bar{\eta} \cdot ({}_Q\langle \bar{\xi}', \bar{\xi} \rangle)^*, \bar{\eta}' \rangle \\ &= {}_R\langle \bar{\eta} \cdot {}_Q\langle \bar{\xi}, \bar{\xi}' \rangle, \bar{\eta}' \rangle \\ &= {}_R\langle \bar{\eta} \otimes_Q \bar{\xi}, \bar{\eta}' \otimes_Q \bar{\xi}' \rangle . \end{aligned}$$

## 3.4 Frobenius reciprocity

LEMMA 3.4.1 *If* $\sigma, \sigma' \in Hom({}_PX_Q, {}_PY_Q)$, *if* ${}_PY_Q$ *is an irreducible bimodule - meaning that* $End({}_PY_Q) = \mathbb{C}\ id_Y$ *- and if* $\{\xi_i\}_i$ *and* $\{\eta_j\}_j$ *are left-bases for* ${}_PX$ *and* ${}_PY$ *respectively, we have:*

$$\sigma\sigma'^* = \alpha\ id_Y, \text{ where } \alpha \in \mathbb{C} \text{ satisfies}$$

$$\alpha = \frac{1}{dim\ {}_PY}\sum_j \langle \sigma\sigma'^*\eta_j, \eta_j\rangle = \frac{1}{dim\ {}_PY}\sum_i \langle \sigma'^*\sigma\xi_i, \xi_i\rangle .$$

*Similarly, if* $\{\xi_i'\}_i$ *and* $\{\eta_j'\}_j$ *are right-bases for* $X_Q$ *and* $Y_Q$ *respectively, we have:*

$$\alpha = \frac{1}{dim\ Y_Q}\sum_j \langle \sigma\sigma'^*\eta_j', \eta_j'\rangle = \frac{1}{dim\ Y_Q}\sum_i \langle \sigma'^*\sigma\xi_i', \xi_i'\rangle .$$

**Proof:** Begin by noting that

$$\begin{aligned}
\sum_j \langle \sigma\sigma'^*\eta_j, \eta_j\rangle &= \sum_j tr_P\ ({}_P\langle \sigma'^*\eta_j, \sigma^*\eta_j\rangle\ ) \\
&= \sum_j\sum_i tr_P\ ({}_P\langle \sigma'^*\eta_j, \xi_i\rangle\ {}_P\langle \xi_i, \sigma^*\eta_j\rangle\ ) \\
&= \sum_j\sum_i tr_P\ (\ {}_P\langle \xi_i, \sigma^*\eta_j\rangle\ {}_P\langle \sigma'^*\eta_j, \xi_i\rangle) \\
&= \sum_j\sum_i tr_P\ (\ {}_P\langle \sigma\xi_i, \eta_j\rangle\ {}_P\langle \eta_j, \sigma'\xi_i\rangle) \\
&= \sum_i tr_P\ (\ {}_P\langle \sigma\xi_i, \sigma'\xi_i\rangle) \\
&= \sum_i \langle \sigma\xi_i, \sigma'\xi_i\rangle \\
&= \sum_i \langle \sigma'^*\sigma\xi_i, \xi_i\rangle\ ;
\end{aligned}$$

on the other hand, it follows from equation (3.2.6) that

$$\sum_j \langle \sigma\sigma'^*\eta_j, \eta_j\rangle = \sum_j \alpha||\eta_j||^2 = \alpha\ dim\ {}_PY\ .$$

The right-statement has an identical proof. □

DEFINITION 3.4.2 *In the notation of the preceding lemma, we write*

$$\langle \sigma, \sigma' \rangle \;=\; \alpha$$

*and note that this defines an inner product on* $Hom({}_PX_Q, {}_PY_Q)$*, whenever* ${}_PY_Q$ *is irreducible.*

We shall assume throughout the rest of this section that ${}_PX_Q, {}_QY_R$ and ${}_PZ_R$ are bimodules of finite type (where $P, Q, R$ are $II_1$ factors).

To start with, deduce from §3.1 (xv)(b) that

$$(\ \pi_l(\xi)\ )(\eta) = \xi \otimes_N \eta = (\ \pi_r(\eta)\ )(\xi)$$

define maps $\pi_l : X^0 \to Hom(Y_R, (X \otimes_Q Y)_R)$ and $\pi_r : Y^0 \to Hom({}_PX, {}_P(X \otimes_Q Y))$.

Also, we shall use the symbols $\xi, \eta, \zeta$ in this section to always denote elements of $X^0, Y^0, Z^0$.

Suppose now that $\sigma \in Hom(\ {}_P(X \otimes_Q Y)_R, {}_PZ_R)$; it may then be verified that the equation

$$(\ F_{(12)}(\sigma)\ )(\zeta \otimes_R \bar{\eta}) \;=\; \sqrt{\frac{dim(X_Q)}{dim(Z_R)}}\ \pi_r(\eta)^*(\sigma^*(\zeta))$$

defines an operator $(\ F_{(12)}(\sigma)\ ) \in Hom(\ {}_P(Z \otimes_R \bar{Y}), {}_PX)$, which satisfies

$${}_P\langle(\ F_{(12)}(\sigma)\ )(\zeta \otimes_R \bar{\eta}), \xi\rangle \;=\; \sqrt{\frac{dim(X_Q)}{dim(Z_R)}}\ {}_P\langle\zeta, \sigma(\xi \otimes_Q \eta)\rangle\ .$$

Two applications of the above identity show that, if $q \in Q$, then

$$\begin{aligned}
&{}_P\langle(\ F_{(12)}(\sigma)\ )(\ (\zeta \otimes_R \bar{\eta}) \cdot q\ ), \xi\rangle \\
&= {}_P\langle(\ F_{(12)}(\sigma)\ )(\zeta \otimes_R \overline{q^* \cdot \eta}), \xi\rangle \\
&= \sqrt{\frac{dim(X_Q)}{dim(Z_R)}}\ {}_P\langle\zeta, \sigma(\xi \otimes_Q q^* \cdot \eta)\rangle \\
&= \sqrt{\frac{dim(X_Q)}{dim(Z_R)}}\ {}_P\langle\zeta, \sigma(\xi \cdot q^* \otimes_Q \eta)\rangle \\
&= {}_P\langle(\ F_{(12)}(\sigma)\ )(\zeta \otimes_R \bar{\eta}), \xi \cdot q^*\rangle \\
&= {}_P\langle(\ (\ F_{(12)}(\sigma)\ )(\zeta \otimes_R \bar{\eta})\ ) \cdot q\ , \xi\rangle\ ,
\end{aligned}$$

thereby showing that $F_{(12)}(\sigma)$ is also right-$Q$-linear. Thus,

$$F_{(12)} : Hom(\ _P(X \otimes_Q Y)_R, {}_P Z_R) \ \to Hom(\ _P(Z \otimes_R \bar{Y})_Q, {}_P X_Q) \ .$$

Similarly, it is seen that the equation

$$(\ F_{(23)}(\sigma)\ )(\bar{\xi} \otimes_Q \zeta) \ = \ \sqrt{\frac{dim({}_Q Y)}{dim({}_P Z)}}\ \pi_l(\xi)^*(\sigma^*(\zeta))$$

defines a mapping

$$F_{(23)} : Hom(\ _P(X \otimes_Q Y)_R, {}_P Z_R) \ \to Hom(\ _Q(\bar{X} \otimes_P Z)_R, {}_Q Y_R) \ .$$

As in the discussion at the end of §2, we shall think of $F_{(12)}, F_{(23)}$ as having been defined for all triples of irreducible bimodules of finite type, and we shall write such symbols as $F_{(12)}F_{(23)}$ to denote the composite of these maps defined on appropriate spaces of intertwiners.

PROPOSITION 3.4.3 *With the foregoing notation, and with respect to the inner-product defined in Definition 3.4.2, we have:*

*(a)* $F_{(12)}$ *and* $F_{(23)}$ *are anti-unitary operators;*
*(b)* $F^2_{(12)} \ = \ F^2_{(23)} \ = \ id$ ;
*(c)* $(F_{(12)}F_{(23)}F_{(12)})^2 \ = \ id$ .

**Proof:** (a) and (b): $F_{(12)}$ and $F_{(23)}$ are clearly conjugate-linear. Also, it is seen that if $\xi \in X^0, \eta \in Y^0, \zeta \in Z^0$, then

$$\begin{aligned}\langle F^2_{(12)}(\sigma)(\xi \otimes_Q \eta), \zeta\rangle \ &= \ \sqrt{\frac{dim(Z_R)}{dim(X_Q)}}\ {}_P\langle \bar{\xi}, F_{(12)}(\sigma)(\bar{\zeta} \otimes_Q \eta)\rangle \\ &= \ \langle \xi \otimes_Q \eta, \zeta\rangle \ ,\end{aligned}$$

and hence (b) holds (for $F_{(12)}$). So (a) will be established (for $F_{(12)}$) once we verify that $F_{(12)}$ preserves norms.

For this, let us fix some choice of right-bases $\{\xi_i\}_i, \{\eta_j\}_j$ for $X_Q$ and $Y_R$ respectively, as well as a left-basis $\{\zeta_k\}_k$ for ${}_P Z$.

Then $\{\zeta_k \otimes_R \bar{\eta}_j\}_{k,j}$ is a left-basis for $_P(Z \otimes_R \bar{Y})$; hence, by Lemma 3.4.1, we have, for $\sigma \in Hom(_P(X \otimes_Q Y)_R, {}_PZ_R)$,

$$\begin{aligned}
&(dim\ X_Q)\,||F_{(12)}(\sigma)||^2 \\
&= \sum_i ||(\ F_{(12)}(\sigma)\ )^*\xi_i||^2 \\
&= \sum_i tr_P\ (\ _P\langle (F_{(12)}(\sigma))^*\xi_i, (F_{(12)}(\sigma))^*\xi_i\rangle\ ) \\
&= \sum_{i,j,k} tr_P\ (\ _P\langle F_{(12)}(\sigma)^*\xi_i, \zeta_k \otimes_R \bar{\eta}_j\rangle\ {}_P\langle \zeta_k \otimes_R \bar{\eta}_j, F_{(12)}(\sigma)^*\xi_i\rangle) \\
&= \sum_{i,j,k} tr_P\ (\ _P\langle \xi_i, F_{(12)}(\sigma)(\zeta_k \otimes_R \bar{\eta}_j)\rangle\ {}_P\langle F_{(12)}(\sigma)(\zeta_k \otimes_R \bar{\eta}_j), \xi_i\rangle) \\
&= \frac{dim\ X_Q}{dim\ Z_R} \sum_{i,j,k} tr_P\ (\ _P\langle \sigma(\xi_i \otimes_Q \eta_j), \zeta_k\rangle\ {}_P\langle \zeta_k, \sigma(\xi_i \otimes_Q \eta_j)\rangle) \\
&= \frac{dim\ X_Q}{dim\ Z_R} \sum_{i,j} tr_P\ (\ _P\langle \sigma(\xi_i \otimes_Q \eta_j), \sigma(\xi_i \otimes_Q \eta_j)\rangle) \\
&= \frac{dim\ X_Q}{dim\ Z_R} \sum_{i,j} \langle \sigma(\xi_i \otimes_Q \eta_j), \sigma(\xi_i \otimes_Q \eta_j)\rangle \\
&= dim\ X_Q\ ||\sigma||^2\ .
\end{aligned}$$

An essentially identical argument disposes of (a) and (b) for $F_{(23)}$ as well.

(c) If $\sigma \in Hom(_PA_Q, {}_PB_Q)$, it is easily verified that the equation

$$\bar{\sigma}(\bar{\xi}) = \overline{\sigma(\xi)}\ ,\ \xi \in A$$

defines an operator $\bar{\xi} \in Hom(_Q\bar{A}_P, {}_Q\bar{B}_P)$ and that the assignment $\sigma \mapsto \bar{\sigma}$ defines an anti-unitary operator.

A direct computation shows that

$$(F_{(12)}F_{(23)}F_{(12)})(\sigma) = K\ \bar{\sigma}\ ,$$

for a certain positive constant $K$; it fcllows that if there exists even one non-zero intertwiner $\sigma \in Hom(_P(X \otimes_Q Y)_R, {}_PZ_R)$, then

$$K = \frac{||(F_{(12)}F_{(23)}F_{(12)})(\sigma)||}{||\bar{\sigma}||} = 1\ ;$$

thus, we have

$$(F_{(12)}F_{(23)}F_{(12)})(\sigma) \;=\; \bar{\sigma}\;. \tag{3.4.10}$$

An identical argument shows that also

$$(F_{(23)}F_{(12)}F_{(23)})(\sigma) \;=\; \bar{\sigma}\;;$$

the last two equations - and the obvious fact that $\bar{\bar{\sigma}} = \sigma$ - establish (c) and the proposition.

□

As in §2, it follows from the above proposition that we have an $S_3$-action on spaces of intertwiners; the proof of the proposition shows, for instance, that $F_{(13)}(\xi) = \bar{\xi}$. We proceed now towards something like quantum 6j symbols that can be constructed from bimodules.

Thus, suppose that ${}_PX_Q, {}_QA_R, {}_RY_S, {}_PD_S, {}_PB_R, {}_QC_S$ are irreducible bimodules of finite type, and $\xi_1 \in \mathcal{H}^B_{XA}, \xi_2 \in \mathcal{H}^D_{XC}, \xi_3 \in \mathcal{H}^D_{BY}, \xi_4 \in \mathcal{H}^C_{AY}$, where we write $\mathcal{H}^B_{XA} = Hom({}_P(X \otimes_Q A)_R, {}_PB_R)$. We then have three essentially equivalent ways of encoding this set of data:

(1) by means of a labelled tetrahedron (with edges labelled by bimodules and faces by intertwiners) - as in figure (2.1);

(2) by means of a rectangular diagram (with 4 bimodules in the corners, one bimodule labelling horizontal maps (right-multiplication), and one bimodule labelling vertical maps (left-multiplication), and one intertwiner labelling each map) - as in Figure 2.2;

(3) by means of the diagram (2.1.2) - with the $N$'s there changed appropriately.

Given such a set of data, since $End({}_PD_S) = \mathbb{C}id_D$ for the irreducible bimodule ${}_PD_S$, we may define

$$W(\xi_1,\xi_2,\xi_3,\xi_4)\; id_D \;=\; \xi_3 \circ (\xi_1 \otimes_R id_Y) \circ (id_X \otimes_Q \xi_4^*) \circ \xi_2^*\;; \tag{3.4.11}$$

$$Z(\xi_1,\xi_2,\xi_3,\xi_4) \;=\; \sqrt{\frac{1}{d_r(B)d_l(C)}}\,W(\xi_1,\xi_2,\xi_3,\xi_4)\;, \tag{3.4.12}$$

where we write $d_l(\cdot)$ (resp., $d_r(\cdot)$) for the left-dimension (resp., right-dimension) of a bimodule.

In order to verify tetrahedral symmetry for the '6j symbol' thus obtained, we only need to check 'invariance' under a set of generators for $S_4$. We shall do this for the three generators $(14)(23), (34), (23)$.

Let $\phi \in \mathcal{S}_1(T)$ be the 'state' of the tetrahedron which corresponds to the above-mentioned data (and the ordering corresponding to the identity permutation). Then, it is seen that

$$\begin{array}{rcl} Z(\ ((14)(23)) \cdot \phi) & = & Z(\bar{\xi_4}, \bar{\xi_3}, \bar{\xi_2}, \bar{\xi_1}) \\ Z(\ (23) \cdot \phi) & = & Z(F_{(12)}(\xi_1), \xi_3, \xi_2, F_{(23)}(\xi_4)) \\ Z(\ (34) \cdot \phi) & = & Z(F_{(23)}(\xi_1), F_{(23)}(\xi_2), \xi_4, \xi_3) \end{array} \tag{3.4.13}$$

LEMMA 3.4.4 *(Invariance under* $(14)(23)$*)*
*With* $A, B, C, D, X, Y$ *and* $\xi_1, \xi_2, \xi_3, \xi_4$ *as above, we have*

$$Z(\xi_1, \xi_2, \xi_3, \xi_4) = Z(\bar{\xi_4}, \bar{\xi_3}, \bar{\xi_2}, \bar{\xi_1}) .$$

**Proof:** Consider the following diagram of bimodules:

$$\begin{array}{ccccccc} X \otimes A \otimes Y & & & \stackrel{id_X \otimes \xi_4}{\rightarrow} & & & X \otimes C \\ & \searrow J & & & & \swarrow J & \\ & & \bar{Y} \otimes \bar{A} \otimes \bar{X} & \stackrel{\bar{\xi_4} \otimes id_{\bar{X}}}{\rightarrow} & \bar{C} \otimes \bar{X} & & \\ \xi_1 \otimes id_Y \downarrow & & id_{\bar{Y}} \otimes \bar{\xi_1} \downarrow & & \downarrow \bar{\xi_2} & & \downarrow \xi_2 \\ & & \bar{Y} \otimes \bar{B} & \stackrel{\bar{\xi_3}}{\rightarrow} & \bar{D} & & \\ & \swarrow J & & & & \searrow J & \\ B \otimes Y & & & \stackrel{\xi_3}{\rightarrow} & & & D \end{array} \tag{3.4.14}$$

where we have simply written $J$ to denote the anti-unitary map $\xi \mapsto \bar{\xi}$ from any bimodule to its contragredient. (We have also omitted subscripts for tensor-products, in the interest of typographical convenience.)

It follows from Remark 3.3.2 and Proposition 3.3.6 that the squares in the above diagram which involve two $J$ maps are commutative (basically, since $\bar{\xi} = J\xi J$).

Hence it follows that

$$
\begin{aligned}
& Z(\xi_1, \xi_2, \xi_3, \xi_4)\ id_D \\
&= \sqrt{\frac{1}{d_r(B)d_l(C)}}\ \xi_3 \circ (\xi_1 \otimes_R id_Y) \circ (id_X \otimes_Q \xi_4^*) \circ \xi_2^* \\
&= \sqrt{\frac{1}{d_r(B)d_l(C)}} (J\bar{\xi}_3 J^*) \circ (J(id_{\bar{Y}} \otimes_R \bar{\xi}_1)J^*) \circ \\
&\qquad \circ (J(\bar{\xi}_4^* \otimes_Q id_{\bar{X}})J^*) \circ (J\bar{\xi}_2^* J^*) \\
&= \sqrt{\frac{1}{d_l(\bar{B})d_r(\bar{C})}} J(\bar{\xi}_3 \circ (id_{\bar{Y}} \otimes_R \bar{\xi}_1) \circ (\bar{\xi}_4^* \otimes_Q id_{\bar{X}}) \circ \bar{\xi}_2^*)J^* \\
&= \sqrt{\frac{1}{d_r(\bar{C})d_l(\bar{B})}} J(\bar{\xi}_2 \circ (\bar{\xi}_4 \otimes_Q id_{\bar{X}}) \circ (id_{\bar{Y}} \otimes_R \bar{\xi}_1{}^*) \circ \bar{\xi}_3^*)^* J^* \\
&= J(\overline{Z(\bar{\xi}_4, \bar{\xi}_3, \bar{\xi}_2, \bar{\xi}_1)}\ id_{\bar{D}})J^* \\
&= Z(\bar{\xi}_4, \bar{\xi}_3, \bar{\xi}_2, \bar{\xi}_1)\ id_D\ ,
\end{aligned}
$$

as desired. □

LEMMA 3.4.5 *(Invariance under* (23)*)*
*With* $A, B, C, D, X, Y$ *and* $\xi_1, \xi_2, \xi_3, \xi_4$ *as above, we have*

$$
Z(\xi_1, \xi_2, \xi_3, \xi_4) \ = \ \overline{Z(F_{(12)}(\xi_1), \xi_3, \xi_2, F_{(23)}(\xi_4))}\ .
$$

**Proof:** By definition, we have:

$$
\begin{aligned}
& \sqrt{d_r(X)d_l(Y)}\ Z(F_{(12)}(\xi_1), \xi_3, \xi_2, F_{(23)}(\xi_4))\ id_D \\
&\quad = \ (\xi_2 \circ (F_{(12)}(\xi_1) \otimes_Q id_C) \circ (id_B \otimes_R (F_{(23)}(\xi_4))^*) \circ \xi_3^*)
\end{aligned}
$$

while

$$
\begin{aligned}
& \sqrt{d_r(B)d_l(C)}\ \overline{Z(\xi_1, \xi_2, \xi_3, \xi_4)}\ id_D \\
&\quad = \ (\ (\xi_3 \circ (\xi_1 \otimes_R id_Y) \circ (id_X \otimes_Q \xi_4^*) \circ \xi_2^*)\ )^* \\
&\quad = \ \xi_2 \circ (id_X \otimes_Q \xi_4) \circ (\xi_1^* \otimes_R id_Y) \circ \xi_3^*\ .
\end{aligned}
$$

Hence it suffices to show that

$$
\begin{aligned}
& (F_{(12)}(\xi_1) \otimes_Q id_C) \circ (id_B \otimes_R (F_{(23)}(\xi_4)^*) \\
&\quad = \ \sqrt{\frac{d_r(X)d_l(Y)}{d_r(B)d_l(C)}} (id_X \otimes_Q \xi_4) \circ (\xi_1^* \otimes_R id_Y)\ .
\end{aligned}
$$

Fix a left-basis $\{\alpha_i\}_i$ for $A$; then, for $\beta \in B^0, \eta \in Y^0$, we see that

$$
\begin{aligned}
&( \ (F_{(12)}(\xi_1) \otimes_Q id_C) \circ (id_B \otimes_R (F_{(23)}(\xi_4)^*) \ )(\beta \otimes_R \eta) \\
&= \sqrt{\frac{d_l(Y)}{d_l(C)}} (F_{(12)}(\xi_1) \otimes_Q id_C) \left( \sum_i \beta \otimes_R \bar{\alpha}_i \otimes_Q \xi_4(\alpha_i \otimes_R \eta) \right) \\
&= \sqrt{\frac{d_r(X) d_l(Y)}{d_r(B) d_l(C)}} \sum_i \pi_r(\alpha_i)^* \xi_1^*(\beta) \ \otimes_Q \ \xi_4(\alpha_i \otimes_R \eta) \\
&= \sqrt{\frac{d_r(X) d_l(Y)}{d_r(B) d_l(C)}} \sum_i ( \ (id_X \otimes_Q \xi_4) \pi_r(\alpha_i) \pi_r(\alpha_i)^* \xi_1^*(\beta) \ ) \otimes_R \eta \\
&= \sqrt{\frac{d_r(X) d_l(Y)}{d_r(B) d_l(C)}} ( \ (id_X \otimes_Q \xi_4) \xi_1^*(\beta) \ ) \otimes_R \eta \\
&= \sqrt{\frac{d_r(X) d_l(Y)}{d_r(B) d_l(C)}} ( \ (id_X \otimes_Q \xi_4)(\xi_1^* \otimes_R id_Y) \ ) \ (\beta \otimes_R \eta) \ ,
\end{aligned}
$$

where the steps are justified thus.

The first equality is a consequence of the assertion that

$$
(F_{(23)}(\xi_4)^*)(\eta) \ = \ \sqrt{\frac{d_l(Y)}{d_l(C)}} \left( \sum_i \ \bar{\alpha}_i \otimes_Q \xi_4(\alpha_i \otimes_R \eta) \right) \ ;
$$

to see this, fix $\alpha \in A^0, \gamma \in C^0$, and note that

$$
\begin{aligned}
&\langle \bar{\alpha} \otimes_Q \gamma, \sqrt{\frac{d_l(Y)}{d_l(C)}} \left( \sum_i \ \bar{\alpha}_i \otimes_Q \xi_4(\alpha_i \otimes_R \eta) \right) \rangle_S \\
&= \sqrt{\frac{d_l(Y)}{d_l(C)}} \sum_i \ \langle \bar{\alpha} \otimes_R \gamma, \bar{\alpha}_i \otimes_Q \xi_4(\alpha_i \otimes_R \eta) \rangle_S \\
&= \sqrt{\frac{d_l(Y)}{d_l(C)}} \sum_i \ \langle \gamma, \langle \bar{\alpha}, \bar{\alpha}_i \rangle_Q \cdot \xi_4(\alpha_i \otimes_R \eta) \rangle_S \\
&= \sqrt{\frac{d_l(Y)}{d_l(C)}} \sum_i \ \langle \gamma, {}_Q\langle \alpha, \alpha_i \rangle \xi_4(\alpha_i \otimes_R \eta) \rangle_S
\end{aligned}
$$

$$
\begin{aligned}
&= \sqrt{\frac{d_l(Y)}{d_l(C)}} \langle \gamma, \xi_4( (\sum_i {}_Q\langle \alpha, \alpha_i\rangle \cdot \alpha_i) \otimes_R \eta)\rangle_S \\
&= \sqrt{\frac{d_l(Y)}{d_l(C)}} \langle \gamma, \xi_4(\alpha \otimes_R \eta)\rangle_S \\
&= \langle \sqrt{\frac{d_l(Y)}{d_l(C)}} \pi_l(\alpha)^* \xi_4^*(\gamma), \eta\rangle_S \\
&= \langle (F_{(23)}(\xi_4))(\bar{\alpha} \otimes_Q \gamma), \eta\rangle_S \\
&= \langle \bar{\alpha} \otimes_Q \gamma, (F_{(23)}(\xi_4))^*(\eta)\rangle_S
\end{aligned}
$$

and the assertion follows.

The second equality follows from the definition:

$$
(F_{(12)}(\xi_1))(\beta \otimes_R \bar{\alpha}) = \sqrt{\frac{d_r(X)}{d_r(B)}} \pi_r(\alpha)^* \xi_1^*(\beta) .
$$

The third equality is just the obvious statement that

$$
\xi \otimes_Q \xi_4(\alpha \otimes_Q \eta) = (id_X \otimes_Q \xi_4)(\pi_r(\alpha)\xi \otimes_R \eta) .
$$

The fourth equality is a consequence of the assertion that if $\{\alpha_i\}_i$ is a left-basis for $A$, then

$$
\sum_i \pi_r(\alpha_i)\pi_r(\alpha_i)^* = id_{X \otimes_Q A} .
$$

For this, first note that if $\xi, \xi' \in X^0$, then

$$
\begin{aligned}
{}_P\langle \pi_r(\alpha_i)^*(\xi \otimes_Q \alpha), \xi'\rangle &= {}_P\langle \xi \otimes_Q \alpha, \xi' \otimes_Q \alpha_i\rangle \\
&= {}_P\langle \xi \cdot {}_Q\langle \alpha, \alpha_i\rangle, \xi'\rangle
\end{aligned}
$$

and hence $\pi_r(\alpha_i)^*(\xi \otimes_Q \alpha) = \xi \cdot {}_Q\langle \alpha, \alpha_i\rangle$.

Now observe that

$$
\begin{aligned}
\xi \otimes_Q \alpha &= \sum_i \xi \otimes_Q {}_Q\langle \alpha, \alpha_i\rangle \cdot \alpha_i \\
&= \sum_i \pi_r(\alpha_i)(\xi \cdot {}_Q\langle \alpha, \alpha_i\rangle) \\
&= \sum_i \pi_r(\alpha_i)\pi_r(\alpha_i)^*(\xi \otimes_Q \alpha) ,
\end{aligned}
$$

thus finishing the justification of step (*4), and the proof of the proposition.

□

LEMMA 3.4.6 *(Invariance under* (34)*)*
*With* $A, B, C, D, X, Y$ *and* $\xi_1, \xi_2, \xi_3, \xi_4$ *as above, we have*

$$Z(F_{(23)}(\xi_1), F_{(23)}(\xi_2), \xi_4, \xi_3) = \sqrt{\frac{d_l(A)d_r(B)}{d_r(A)d_l(B)}}\overline{Z(\xi_1, \xi_2, \xi_3, \xi_4)}.$$

**Proof:** We first assert that

$$\begin{aligned} & F_{(23)}(\xi_3 \circ (\xi_1 \otimes_R id_Y) \circ (id_X \otimes_Q \xi_4^*)) \\ = & \sqrt{\frac{d_l(B)d_l(C)}{d_l(A)d_l(D)}}\ \xi_4 \circ (F_{(23)}(\xi_1) \otimes_R id_Y) \circ (id_{\bar{X}} \otimes_P \xi_3^*)\ . \end{aligned} \tag{3.4.15}$$

To see this, fix arbitrary $\xi \in X^0, \gamma \in C^0, \delta \in D^0$; then we may write

$$\begin{aligned} \xi_4^*(\gamma) &= \sum_i \alpha_i \otimes_R \eta_i \ \in \ A \otimes_R Y \\ \xi_3^*(\delta) &= \sum_j \beta_j \otimes_R \eta'_j \ \in \ B \otimes_R Y \end{aligned}$$

for appropriate $\alpha_i \in A^0, \beta_j \in B^0, \eta_i, \eta'_j \in Y^0$. Then we see that

$$\begin{aligned} & \sqrt{\frac{d_l(D)d_l(A)}{d_l(C)d_l(B)}}\ \langle F_{(23)}(\xi_3 \circ (\xi_1 \otimes_R id_Y) \circ (id_X \otimes_Q \xi_4^*))(\bar{\xi} \otimes_P \delta), \gamma\rangle \\ = & \sqrt{\frac{d_l(A)}{d_l(B)}}\langle \delta, \xi_3 \circ (\xi_1 \otimes_R id_Y) \circ (id_X \otimes_Q \xi_4^*)\ (\xi \otimes_Q \gamma)\rangle \\ = & \sqrt{\frac{d_l(A)}{d_l(B)}}\sum_{i,j}\langle \beta_j \otimes_R \eta'_j, (\xi_1 \otimes_R id_Y)(\xi \otimes_Q \alpha_i \otimes_R \eta_i)\rangle \\ = & \sqrt{\frac{d_l(A)}{d_l(B)}}\sum_{i,j} tr_{P\ P}\langle \beta_j \otimes_R \eta'_j, \xi_1(\xi \otimes_Q \alpha_i) \otimes_R \eta_i\rangle \end{aligned}$$

$$
\begin{aligned}
&= \sqrt{\frac{d_l(A)}{d_l(B)}}\, tr_P \left( \sum_{i,j} {}_P\langle \beta_j \cdot {}_R\langle \eta'_j, \eta_i \rangle, \xi_1(\xi \otimes_Q \alpha_i) \rangle \right) \\
&= tr_P \left( \sum_{i,j} {}_P\langle F_{(23)}(\xi_1)\ (\bar{\xi} \otimes_P \beta_j \cdot {}_R\langle \eta'_j, \eta_i \rangle), \alpha_i \rangle \right) \\
&= tr_P \left( \sum_{i,j} {}_P\langle\ (F_{(23)}(\xi_1)(\bar{\xi} \otimes_P \beta_j)\ ) \cdot {}_R\langle \eta'_j, \eta_i \rangle, \alpha_i \rangle \right) \\
&= tr_P \left( \sum_{i,j} {}_P\langle\ (F_{(23)}(\xi_1)(\bar{\xi} \otimes_P \beta_j)\ ) \otimes_R \eta'_j, \alpha_i \otimes_R \eta_i \rangle \right) \\
&= tr_P \left( \sum_{j} {}_P\langle (F_{(23)}(\xi_1) \otimes_R id_Y)(\bar{\xi} \otimes_P \beta_j \otimes_R \eta'_j), \xi_4^*(\gamma) \rangle \right) \\
&= tr_P \left( {}_P\langle \xi_4 (F_{(23)}(\xi_1) \otimes_R id_Y)(id_{\bar{X}} \otimes_P \xi_3^*)(\bar{\xi} \otimes_P \delta), \gamma \rangle \right) \\
&= \langle \xi_4 (F_{(23)}(\xi_1) \otimes_R id_Y)(id_{\bar{X}} \otimes_P \xi_3^*)(\bar{\xi} \otimes_P \delta), \gamma \rangle \ ;
\end{aligned}
$$

since $\xi, \delta, \gamma$ were arbitrary, the proof of the assertion (3.4.15) is complete.

Hence we see that

$$
\begin{aligned}
&Z(F_{(23)}(\xi_1), F_{(23)}(\xi_2), \xi_4, \xi_3) \\
&= \sqrt{\frac{1}{d_r(A)d_l(D)}}\ \xi_4 \circ (F_{(23)}(\xi_1) \otimes_R id_Y) \circ \\
&\qquad\qquad \circ (id_{\bar{X}} \otimes_P \xi_3^*) \circ F_{(23)}(\xi_2)^* \\
&= \sqrt{\frac{1}{d_r(A)d_l(D)}} \sqrt{\frac{d_l(A)d_l(D)}{d_l(B)d_l(C)}} F_{(23)}(\xi_3 \circ (\xi_1 \otimes_R id_Y) \circ \\
&\qquad\qquad \circ (id_X \otimes_Q \xi_4^*)) \circ (F_{(23)}(\xi_2))^* \\
&= \sqrt{\frac{d_l(A)}{d_r(A)}} \sqrt{\frac{1}{d_l(B)d_l(C)}} (\xi_3 \circ (\xi_1 \otimes_R id_Y) \circ (id_X \otimes_Q \xi_4^*) \circ \xi_2^*)^* \\
&= \sqrt{\frac{d_l(A)}{d_r(A)}} \sqrt{\frac{1}{d_l(B)d_l(C)}} \sqrt{d_r(B)d_l(C)}\ \overline{Z(\xi_1, \xi_2, \xi_3, \xi_4)} \\
&= \sqrt{\frac{d_l(A)}{d_r(A)}} \sqrt{\frac{d_r(B)}{d_l(B)}}\ \overline{Z(\xi_1, \xi_2, \xi_3, \xi_4)}\ .
\end{aligned}
$$

Note that, for the third equality, we have used the fact that $F_{(23)}$ is anti-unitary; for if $S, T \in Hom(X \otimes_N Y, Z)$, with $Z$ irreducible, then

$$F_{(23)}(T)F_{(23)}(S)^* = \langle F_{(23)}(T), F_{(23)}(S)\rangle = \overline{\langle T, S\rangle} = \overline{TS^*} . \qquad \square$$

We summarise the findings of this section in the following remark, in the form that will be pertinent for us.

REMARK 3.4.7 Suppose $\mathcal{M}$ is a finite collection of irreducible $N-N$-bimodules of finite type which is (a) closed under contragredients, and (b) 'closed under tensor products' in the sense that $X \otimes_N Y \cong \oplus_{Z \in \mathcal{M}} N_{XY}^Z \cdot Z$ for all $X, Y \in \mathcal{M}$, where we write $n \cdot Z$ to mean a direct sum of $n$ copies of $Z$. Then, it is true that $\mathcal{M}$ is a basis for a finite-dimensional fusion algebra. There is a natural choice of intertwiner spaces, viz. $\mathcal{H}_{XY}^Z = Hom(X \otimes_N Y, Z)$. (We will see, in Remark 3.5.1(e), why this is a finite-dimensional space.) The content of Proposition 3.4.3 is that the Frobenius reciprocity maps $F_\sigma, \sigma \in S_3$ give an $S_3$-action by anti-unitary/unitary maps.

Further, it follows from the uniqueness of the dimension function on a finite-dimensional fusion algebra and the multiplicativity of the (left- as well as the right-) $M$-dimension under tensor products that $d_l(A) = d_r(A)\ \forall A \in \mathcal{M}$. This shows that the $Z$'s defined by equations (3.4.11), when restricted to $\mathcal{M}$, do admit an '$S_4$-action' under which it is 'invariant', in the sense that it changes to its conjugate under the action of any odd permutation. (By Lemma 3.4.4 and Lemma 3.4.5, we already had 'invariance' under the subgroup generated by (14)(23) and (23), with no 'finiteness assumptions'. The foregoing remark about the equality of left- and right-dimensions and Lemma 3.4.6 shows that, in the presence of the finiteness assumption, we also have invariance under (34) and hence under the entire group $S_4$.)

## 3.5 TQFTs from subfactors

Suppose we have an inclusion $N \subset M$ of $II_1$ factors. We shall only consider such inclusions which are *unital*, meaning that the multiplicative identity 1 of $M$ is contained in $N$. The **index** of such an inclusion is defined to be

$$[M : N] \;=\; dim_N(\; {}_NL^2(M)\;)\;. \qquad (3.5.16)$$

In the sequel, we shall reserve the expression **subfactor** to mean a unital inclusion of $II_1$ factors, as above, with finite index. In the rest of this section, we assume that $N \subset M$ is such a subfactor, and we shall reserve the symbol $\alpha$ for the $N - M$-bimodule defined by

$$\alpha \;=\; {}_NL^2(M)_M\;.$$

Thus, by definition,

$$dim_N\;\alpha \;=\; [M : N]\;. \qquad (3.5.17)$$

We list some basic facts concerning subfactors in the following remark.

REMARK 3.5.1 (a) Consideration of the map $x \mapsto x^*$ shows that

$$\overline{{}_M(L^2(M))_M} \;\cong\; {}_ML^2(M)_M\;.$$

By 'restriction', it follows that we have an identification

$$\bar{\alpha} \;=\; {}_M(L^2(M))_N\;.$$

(b) If $X$ is any left $M$-module, then

$${}_NX \;\cong\; {}_N(\alpha \otimes_M X)\;,$$

and, in particular,

$$dim_N\;X \;=\; [M : N]\; dim_M\; X\;.$$

The above statement is true even if $[M : N] = \infty$; in particular, we see that the following conditions on a unital inclusion $N \subset M$ (of $II_1$ factors) are equivalent:

(i) $[M : N] \;<\; \infty$ ;

(ii) there exists a left $M$-module of finite type such that $dim_N\; X < \infty$.

In this case, we have

$$[M : N] \;=\; \frac{dim_N\ X}{dim_M\ X}\ .$$

There are entirely analogous dual statements about right $M$-modules.

(c) The following conditions on a left $M$-module $X$ are equivalent:
(i) $dim_M\ X < \infty$ ;
(ii) $M' = End({}_M X)$ is a $II_1$ factor.
If these equivalent conditions are met, then

$$dim_{M'}\ X \;=\; \frac{1}{dim_M\ X}\ .$$

(d) If $[M : N] < \infty$, then $N' \cap M = \{m \in M : mn = nm\ \forall\ n \in N\}$ is a finite-dimensional $C^*$-algebra; and in fact, $dim_{\mathbb{C}}(N' \cap M) \leq [M : N]$ .

(e) Suppose ${}_P X_Q$ is a bimodule of finite type. Let $N = \pi_l(P)$ and $M = \pi_r(Q)'$. It follows from (c) that $M$ is a $II_1$ factor. Clearly $N \subset M$, and (b) implies that this is a subfactor of finite index; in fact,

$$[M : N] \;=\; \frac{dim_N\ X}{dim_M\ X} \;=\; (dim\ {}_P X)(dim\ X_Q)\ .$$

By definition, we have

$$End({}_P X_Q) = \pi_l(P)' \cap \pi_r(Q)' = N' \cap M\ ,$$

and we thus conclude that:

$End({}_P X_Q)$ is a finite-dimensional $C^*$-algebra, for any finite-type bimodule ${}_P X_Q$. More generally (in fact, equivalently) the space $Hom({}_P X_Q, {}_P Y_Q)$ is finite-dimensional, for any two bimodules of finite type.

(f) Consider the following facts, valid in any finite-dimensional $C^*$-algebra, say $A$ (since such an algebra is isomorphic to a finite direct sum of matrix algebras):

(i) the identity of $A$ is the sum of the minimal projections of its centre; i.e., we have $1 = \sum_{i=1}^{z} q_i$, where $\{q_i : 1 \leq i \leq z\}$ is the set of minimal projections of $Z(A)$;

(ii) each minimal central projection $q_i$ is expressible as a sum of minimal projections in $A$; i.e., $q_i = \sum_{j=1}^{m_i} p_j^i$, where $\{p_j^i : 1 \leq j \leq m_i\}$ is a set of minimal projections in $A$ - in fact, in $q_i A$; further, for any two indices $1 \leq j, k \leq m_i$, there exists an element $u_{jk} \in q_i A$ such that $u_{jk}^* u_{jk} = p_k^i$ and $u_{jk} u_{jk}^* = p_j^i$.

Suppose now that ${}_P X_Q$ is as in (e) above. Let $A = End({}_P X_Q)$, and note that the association $p \leftrightarrow pX = ran(p)$ sets up a bijection between projections in $A$ and $P-Q$-submodules of $X$; further there exists a natural identification of $Hom(pX, p'X)$ with $p'Ap$.

It is now easily verified that the above two facts translate as follows:

(i)′ We have a decomposition $X = \oplus_{i=1}^{z} q_i X$ of $X$ into submodules which are 'disjoint' in the sense that $Hom(q_i X, q_l X) = \{0\}$ if $i \neq l$; and

(ii)′ the $q_i X$ are *isotypical* in the sense that they admit a further decomposition $q_i X = \oplus_{j=1}^{m_i} p_j^i X$ as a direct sum of pairwise isomorphic irreducible bimodules.

To every finite index subfactor $N \subset M$, we naturally associate four collections of bimodules as follows:

Let $\mathcal{M}_{00}$ (resp., $\mathcal{M}_{01}, \mathcal{M}_{10}, \mathcal{M}_{11}$) denote the set of isomorphism classes of irreducible $N-N$ (resp., $N-M$, $M-N$, $M-M$) bimodules that arise as submodules of $(\alpha\bar{\alpha})^n$ (resp., $(\alpha\bar{\alpha})^n\alpha$, $\bar{\alpha}(\alpha\bar{\alpha})^n$, $(\bar{\alpha}\alpha)^n$) for some positive integer $n$. (Here we have dispensed with the 'tensor-product symbol' and write, for instance, $\alpha\bar{\alpha}\alpha$ for $\alpha \otimes_M \bar{\alpha} \otimes_N \alpha$.)

It is easily seen that the collection $\cup_{i,j=0}^{1} \mathcal{M}_{ij}$ is 'closed under contragredients and tensor-products' in the sense that if $X \in \mathcal{M}_{ij}, Y \in \mathcal{M}_{jk}$, then $\bar{X} \in \mathcal{M}_{ji}$ and $X \otimes_{M_j} Y \cong \oplus_{Z \in \mathcal{M}_{ik}} N_{XY}^Z \cdot Z$, where we have written $M_j$ to mean $N$ or $M$ according as $j = 0$ or 1, and the understanding is that $N_{XY}^Z \neq 0$ for at most finitely many $Z$.

It is a fact that $\mathcal{M}_{11}$ is a finite set if and only if $\mathcal{M}_{00}$ is a finite set if and only if $\cup_{i,j=0}^{1}\mathcal{M}_{ij}$ is a finite set. When this condition is satisfied, the subfactor is said to have **finite depth**.

We shall only deal with finite-depth subfactors in the sequel. In this case, it is seen from remark 3.4.7 that both $\mathcal{M}_{00}$ and $\mathcal{M}_{11}$ are bases for finite-dimensional fusion algebras. Henceforth, we fix $\mathcal{M} = \mathcal{M}_{00}$, so that remark 3.4.7 is applicable to this $\mathcal{M}$. To economise on notation, we shall consequently use the symbol $\otimes$ to signify $\otimes_N$ in the rest of this chapter.

As before, we shall be concerned with pictures such as Figure 2.2 or the equivalent diagram 2.1.2, where all the bimodules come from $\mathcal{M}$ and the intertwiners are $N-N$ linear maps between appropriate bimodules. In fact, we shall want to look at larger rectangles ($k \times l$, and not just $1 \times 1$).

We need some terminology. Temporarily fix positive integers $k, l$. By a *marked course of size (k,l)*, we shall mean a tuple of the form $\gamma = (\overline{m}, \overline{n}, \mathbb{X}, \mathbb{Y})$, where:

(i) $\overline{m} = (m_1, m_2, \cdots, m_p)$, where $m_i$ are non-negative integers and $m_i > 0$ for $1 \leq i \leq (p-1)$, and $\sum_i m_i = k$;

(ii) $\overline{n} = (n_1, n_2, \cdots, n_p)$, where $n_j$ are non-negative integers and $n_j > 0$ for $2 \leq i \leq p$, and $\sum_j n_j = l$;

(iii) $\mathbb{X} = (X_1, X_2, \cdots, X_k)$, $X_i \in \mathcal{M}$; and

(iv) $\mathbb{Y} = (Y_1, Y_2, \cdots, Y_l)$, $Y_j \in \mathcal{M}$.

We would like to think of such a 'marked course' as follows: starting at some point in the plane, first walk $n_1$ steps to the right, then $m_1$ steps down, then $n_2$ steps to the right, then $m_2$ steps down, etc., and mark the $i$-th step in the downward direction (resp., the $j$-th step to the right) by the symbol $X_i\cdot$ (resp., $\cdot Y_j$). (The possibility of there being no initial step to the right or final step down is the reason for the peculiar positivity requirements in (i) and (ii) above.)

A marked course will be said to be *tethered* if its 'initial' and 'final' points have been marked by some two elements, say $A$ and $D$, of $\mathcal{M}$. An example of such a 'tethered marked course', with $k = 3, l = 4, p = 3$ and $\overline{m} = (2, 1, 0), \overline{n} = (0, 1, 3)$, is illustrated in Figure 3.1.

A labelling of such a 'tethered marked course' is a tuple $(\sigma, \tau)$, where (i) $\sigma = (\sigma_1, \sigma_2, \cdots, \sigma_k)$, $\sigma_i \in Hom(X_i \otimes s(\sigma_i), r(\sigma_i))$,

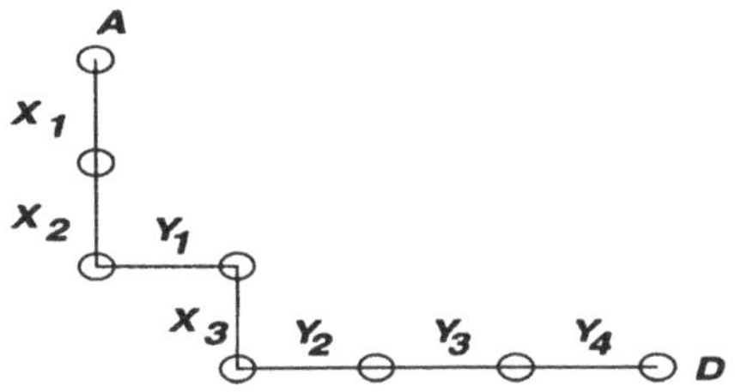


Figure 3.1: A 'tethered marked course'

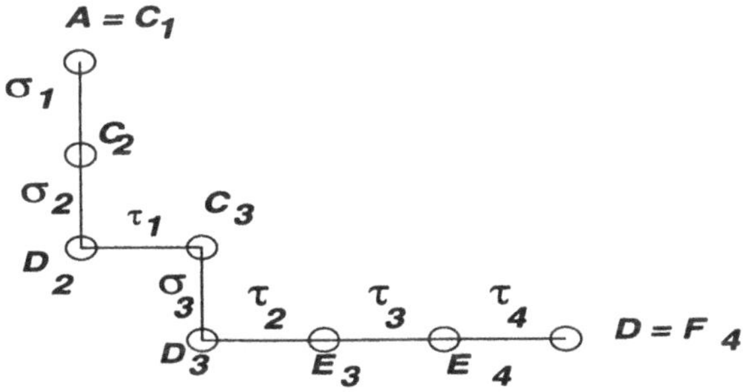


Figure 3.2: A 'labelled tethered marked course'

and $s(\sigma_i), r(\sigma_i) \in \mathcal{M}$ for $1 \leq i \leq k$ (where we think of $s(\sigma_i)$ and $r(\sigma_i)$ as the 'source' and 'range', respectively, of $\sigma_i$); (ii) $\tau = (\tau_1, \tau_2, \cdots, \tau_l)$ , $\tau_j \in Hom(s(\tau_j) \otimes Y_j, r(\tau_j))$, and $s(\tau_j), r(\tau_j) \in \mathcal{M}$ for $1 \leq j \leq l$; and (iii) the bimodules must 'patch up correctly', as in Figure 3.2; thus, for a labelling of the illustrated 'tethered marked course', if we have $s(\sigma_i) = C_i, r(\sigma_i) = D_i$, and $s(\tau_j) = E_j, r(\tau_j) = F_j$, then we must have $A = C_1, D_1 = C_2, D_2 = E_1, F_1 = C_3, D_3 = E_2, F_2 = E_3, F_3 = E_4$ and $F_4 = D$.

Any labelling $(\sigma, \tau)$, as above, yields an $N - N$-linear map

$$\sigma\tau : X_k \otimes \cdots \otimes X_1 \otimes A \otimes Y_1 \otimes \cdots \otimes Y_l \to D ;$$

for instance, in the illustrated example above, this is just the composite map

$$\begin{array}{ll} X_3 \otimes X_2 \otimes X_1 \otimes A \otimes Y_1 \otimes \cdots \otimes Y_4 & \\ \overset{id_{(X_3 \otimes X_2)} \otimes \sigma_1 \otimes id_{(Y_1 \otimes \cdots \otimes Y_4)}}{\to} & X_3 \otimes X_2 \otimes r(\sigma_1) \otimes Y_1 \otimes Y_2 \otimes Y_3 \otimes Y_4 \\ \overset{id_{X_3} \otimes \sigma_2 \otimes id_{(Y_1 \otimes Y_2 \otimes Y_3 \otimes Y_4)}}{\to} & X_3 \otimes r(\sigma_2) \otimes Y_1 \otimes Y_2 \otimes Y_3 \otimes Y_4 \\ \overset{id_{X_3} \otimes \tau_1 \otimes id_{(Y_2 \otimes Y_3 \otimes Y_4)}}{\to} & X_3 \otimes r(\tau_1) \otimes Y_2 \otimes Y_3 \otimes Y_4 \\ \overset{\sigma_3 \otimes id_{(Y_2 \otimes Y_3 \otimes Y_4)}}{\to} & r(\sigma_3) \otimes Y_2 \otimes Y_3 \otimes Y_4 \\ \overset{\tau_2 \otimes id_{Y_3 \otimes Y_4}}{\to} & r(\tau_2) \otimes Y_3 \otimes Y_4 \\ \overset{\tau_3 \otimes id_{Y_4}}{\to} & r(\tau_3) \otimes Y_4 \\ \overset{\tau_4}{\to} & D . \end{array}$$

In this example, we shall call $(A, C_2, E_1, C_3, E_2, E_3, E_4, D)$ the *profile* of $(\sigma, \tau)$. In general we shall refer to a tuple $\mathbb{A} = (A = A_0, A_1, \cdots, A_{k+l} = D)$ of bimodules as a 'profile of length $(k + l)$ with source $A$ and range $D$'.

Given a profile $\mathbb{A}$ as above, and a marked course $\gamma$ tethered at $A$ and $D$, then the $q$-th step of $\gamma$ is marked with an $(X_{i_q} \cdot)$ or a $(\cdot Y_{j_q})$ according as the step is in the downward or eastward direction; and the label $(\sigma, \tau)$ specifies an intertwiner $\sigma_{i_q} \in Hom(X_{i_q} \otimes A_{q-1}, A_q)$ or $\tau_{j_q} \in Hom(A_{q-1} \otimes Y_{j_q}, A_q)$ as the case may be; let us refer to the corresponding 'Hom' space as the $q$-th intertwiner space for $\mathbb{A}$, and denote it by $\mathcal{H}(\gamma; q, \mathbb{A})$.

Suppose now that, for each profile $\mathbb{A}$ as above, and for each $1 \leq q \leq (k+l)$, we arbitrarily choose and fix some orthonormal basis $\mathcal{B}_{(\gamma;q,\mathbb{A})}$ for $\mathcal{H}(\gamma; q, \mathbb{A})$. (Thus, $\mathcal{B}_{(\gamma;q,\mathbb{A})}$ is just a maximal collection $\{\xi_r\}_r$ of $N-N$-linear coisometries in $\mathcal{H}(\gamma; q, \mathbb{A})$ such that $\xi_r \xi_s^* = \delta_{rs}$.) Then let us write $\mathcal{B}_\gamma(\mathbb{A})$ for the collection of all labellings $(\sigma, \tau)$ of $\gamma$ which satisfy: (i) $(\sigma, \tau)$ has profile $\mathbb{A}$; and (ii) each of the $\sigma_i$'s and $\tau_j$'s belong to the orthonormal basis for the corresponding intertwiner space. Note that the same intertwiner space $Hom(B \otimes C, D)$ might arise at different steps of $\gamma$ and the corresponding bases are not required to be the same. Also, note that to each collection $\{\mathcal{B}_{(\gamma;q,\mathbb{A})}\}$ of orthonormal bases, the collection $\mathcal{B}_\gamma = \coprod_{\mathbb{A}} \mathcal{B}_\gamma(\mathbb{A})$ is clearly an orthonormal basis for the big intertwiner space $\mathcal{H}_\gamma = Hom(\ (X_k \otimes \cdots \otimes X_1) \otimes A \otimes (Y_1 \otimes \cdots \otimes Y_l), D)$; thus we are using the symbol $\mathcal{B}_\gamma$ to denote any member of a suitable collection of orthonormal bases (with good local properties) of the grand intertwiner space associated to $\gamma$.

Of special interest to us will be the two 'extreme courses' given by $\gamma = ((k,0),(0,l),\mathbb{X},\mathbb{Y})$ and $\gamma' = ((k),(l),\mathbb{X},\mathbb{Y})$. For any string $\sigma$ of length $k$ which is 'consistent with left-tensoring by $\mathbb{X}$', i.e., $\sigma_i \in Hom(X_i \otimes s(\sigma_i), r(\sigma_i))$, we define

$$\begin{aligned} L_\sigma &= \sigma_k \circ (id_{X_k} \otimes \sigma_{k-1}) \circ \cdots \circ (id_{(X_k \otimes \cdots \otimes X_2)} \otimes \sigma_1) \\ &\in Hom(X_k \otimes \cdots \otimes X_1 \otimes s(\sigma), r(\sigma)) \end{aligned} \tag{3.5.18}$$

where we write $s(\sigma) = s(\sigma_1), r(\sigma) = r(\sigma_k)$. Similarly, for any string $\tau$ of length $l$ which is 'consistent with right-tensoring by $\mathbb{Y}$', i.e., $\tau_j \in Hom(s(\tau_j) \otimes Y_j, r(\tau_j))$, we define

$$\begin{aligned} R_\tau &= \tau_l \circ (\tau_{l-1} \otimes id_{Y_l}) \circ \cdots \circ (\tau_1 \otimes id_{(Y_2 \otimes \cdots \otimes Y_l)}) \\ &\in Hom(s(\tau) \otimes Y_1 \otimes \cdots \otimes Y_l, r(\tau)) . \end{aligned} \tag{3.5.19}$$

Then, if $\sigma\tau \in \mathcal{B}_\gamma, \sigma'\tau' \in \mathcal{B}_{\gamma'}$, we have:

$$\begin{aligned} \sigma\tau &= R_\tau \circ (L_\sigma \otimes id_{(Y_1 \otimes \cdots \otimes Y_l)}) \\ \sigma'\tau' &= L_{\sigma'} \circ (id_{(X_k \otimes \cdots \otimes X_1)} \otimes R_{\tau'}) . \end{aligned}$$

Two marked tethered courses, say

$$\begin{aligned} \gamma &= (\overline{m}, \overline{n}, \mathbb{X}, \mathbb{Y}) \\ \gamma' &= (\overline{m}', \overline{n}', \mathbb{X}, \mathbb{Y}) , \end{aligned}$$

will be said to be *comparable* if they have the same (i) size, (ii) 'marking strings' $\mathbb{X}$ and $\mathbb{Y}$ of bimodules, and (iii) 'starting' and 'finishing' points. For any such comparable pair, we then define the $\mathcal{B}_\gamma \times \mathcal{B}_{\gamma'}$ matrix $W^\gamma_{\gamma'}$ with $(\sigma\tau, \sigma'\tau')$-entry given by

$$W^{\sigma\tau}_{\sigma'\tau'} = \langle \sigma\tau, \sigma'\tau' \rangle = \sigma\tau(\sigma'\tau')^* \ . \tag{3.5.20}$$

For convenience of reference, we record various straightforward facts about the '$W$-matrices", so defined, in the following remark.

REMARK 3.5.2 (a) The matrix $W^\gamma_{\gamma'} = ((W^{\sigma\tau}_{\sigma'\tau'}))$ is a unitary matrix, for every pair of 'comparable' marked tethered courses $\gamma, \gamma'$. (This is because it is the matrix of inner products between members of two different orthonormal bases of the same Hilbert space.)

(b) If $\gamma_i, 1 \leq i \leq 3$ is a collection of three pairwise comparable marked tethered courses, then we have

$$W^{\gamma_1}_{\gamma_3} = W^{\gamma_1}_{\gamma_2} W^{\gamma_2}_{\gamma_3} \ ,$$

thanks to the basic Parseval's identity in Hilbert space.

(c) Let $\gamma = ((1,0),(0,1),(X),(Y)), \gamma' = ((1),(1),(X),(Y))$ (thus, $k = l = 1$); and suppose $\sigma = (\xi_1)$, $\tau = (\xi_3)$, $\sigma' = (\xi_2)$, $\tau' = (\xi_4)$, with $s(\xi_1) = s(\xi_4) = A, r(\xi_1) = s(\xi_3) = B, r(\xi_4) = s(\xi_2) = C, r(\xi_3) = r(\xi_2) = D$. Then $W^{\sigma\tau}_{\sigma'\tau'}$ is just the quantity defined as $W(\xi_1, \xi_2, \xi_3, \xi_4)$ in equation (3.4.11); thus the new $W$-matrix is a generalisation of the old one.

(d) It follows from (b) and (c) above that the entry $W^{\sigma\tau}_{\sigma'\tau'}$ of a general large $W$-matrix may be obtained by a sort of 'partition-function-sum' involving the $1 \times 1$ matrices; thus, suppose that $\gamma = ((1,0),(0,2),(X),(Y_1,Y_2))$ and $\gamma' = ((1),(2),(X),(Y_1,Y_2))$, and $\sigma = (\sigma_1), \tau = (\tau_1, \tau_2)$, $\sigma' = (\sigma'_1), \tau = (\tau'_1, \tau'_2)$, then

$$W^{\sigma\tau}_{\sigma'\tau'} = \sum_{\sigma''} W^{\sigma_1\tau_1}_{\sigma''\tau'_1} W^{\sigma''\tau_2}_{\sigma'_1\tau'_2} \ .$$

Suppose $\gamma = ((k,0),(0,l),\mathbb{X},\mathbb{Y})$ and $\gamma' = ((k),(l),\mathbb{X},\mathbb{Y})$ are comparable and start at $A$ and finish at $D$; then we shall use a pictorial depiction of the entry $W^{\sigma\tau}_{\sigma'\tau'}$ of $W^\gamma_{\gamma'}$, as in Figure 3.3.

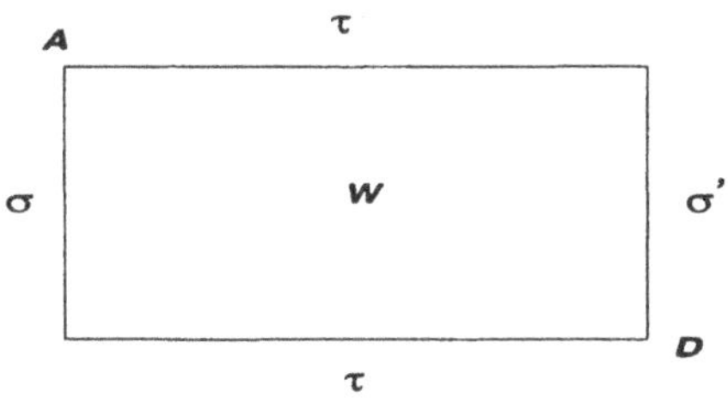


Figure 3.3: $W^{\sigma\tau}_{\sigma'\tau'}$

REMARK 3.5.3 (a) In the sequel, we shall simply write $*$ for the bimodule ${}_N L^2(N)_N$. Further, if $\sigma \in Hom(X \otimes Y, Z)$, we shall use the notation

$$\begin{aligned} \sigma_{\leftrightarrow} &= F_{(12)}(\sigma) \in Hom(Z \otimes \bar{Y}, X) \\ \sigma_{\updownarrow} &= F_{(23)}(\sigma) \in Hom(\bar{X} \otimes Z, Y) \ . \end{aligned}$$

More generally, if $\sigma\tau \in \mathcal{B}_\gamma$ - with $\gamma = ((k,0),(0,l),\mathbb{X},\mathbb{Y})$ - where $\sigma = (\sigma_1, \cdots, \sigma_k)$ and $\tau = (\tau_1, \cdots, \tau_l)$, we shall write

$$\begin{aligned} \tau_{\leftrightarrow} &= (\tau_{l\leftrightarrow}, \cdots \tau_{1\leftrightarrow}) \\ \sigma_{\updownarrow} &= (\sigma_{k\updownarrow}, \cdots \sigma_{1\updownarrow}) \ . \end{aligned}$$

(b) In view of the tetrahedral symmetry enjoyed by the $Z$-function associated to our $W$, we see - on applying each of the permutations (12) and (34) - that, with $\gamma = ((1,0),(0,1),\mathbb{X},\mathbb{Y})$, $\gamma' = ((1),(1),\mathbb{X},\mathbb{Y})$, we have

$$W^{\sigma\tau}_{\sigma'\tau'} = \sqrt{\frac{dim_N(r(\sigma))dim_N(s(\sigma'))}{dim_N(s(\sigma))dim_N(r(\sigma'))}}\ \overline{W^{\sigma'\tau_{\leftrightarrow}}_{\sigma\tau'_{\leftrightarrow}}} \qquad (3.5.21)$$

$$= \sqrt{\frac{dim_N(r(\sigma))dim_N(s(\sigma'))}{dim_N(s(\sigma))dim_N(r(\sigma'))}}\ \overline{W^{\sigma_{\updownarrow}\tau'}_{\sigma'_{\updownarrow}\tau}} \ . \qquad (3.5.22)$$

(In fact this statement is also valid for larger $k, l$; thus suppose $\gamma = ((1,0),(0,1),\mathbb{X},\mathbb{Y})$, $\gamma' = ((1),(1),\mathbb{X},\mathbb{Y})$; then the $W$-value on the left side is a sum of products of the $W$-values of $1 \times 1$ rectangles. Apply the above statement to each of $1 \times 1$ subrectangles; and note that in each product, the contributions

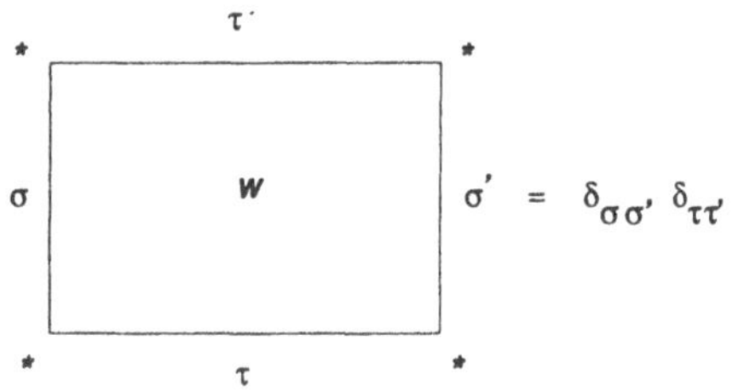


Figure 3.4: Four * flatness

to the square-root terms by each non-corner vertex cancel out due to occurrences in both numerator and denominator; and conclude that the above equations are still valid for this $k \times l$ rectangle.)

All the pieces are now in place for the crucial statement regarding *flatness*.

PROPOSITION 3.5.4 (**Flatness**) *(a) Let* $\gamma = ((k,0),(0,l),\mathbb{X},\mathbb{Y})$ *and* $\gamma' = ((k),(l),\mathbb{X},\mathbb{Y})$, *and suppose* $\sigma\tau \in \mathcal{B}_\gamma$, $\sigma'\tau' \in \mathcal{B}_{\gamma'}$ *satisfy* $s(\sigma) = r(\sigma) = s(\sigma') = r(\sigma') = *$. *Then,*

$$W^{\sigma\tau}_{\sigma'\tau'} = \delta_{\sigma,\sigma'}\delta_{\tau,\tau'} \; .$$

*(b) In the notation of (a), if it is only assumed that* $s(\sigma) = s(\sigma') = *$, *then we have:*

$$W^{\sigma\tau}_{\sigma'\tau'} = \delta_{\sigma,\sigma'} C^{\tau}_{\tau'} \; ;$$

*the above equation says two things: (i) the left side is equal to 0 if* $\sigma \neq \sigma'$; *and (ii) if* $\sigma = \sigma'$, *then the value of the left side depends only on* $r(\sigma)(= s(\tau))$, *and is otherwise independent of the choice of* $\sigma = \sigma'$.

*(c) Fix a* $B \in \mathcal{M}$ *and let* $\sigma$ *be any 'string from* $*$ *to* $B$' *- meaning that* $\sigma = (\sigma_1, \cdots, \sigma_k)$ *where* $\sigma_i \in Hom(X_i \otimes A_i, A_{i+1})$ *for some* $A_i, X_i \in \mathcal{M}$, *subject only to the requirement that* $A_1 = *, A_{k+1} = B$. *Consider the matrix* $C = ((C^{\tau}_{\tau'}))$ *with rows (resp., columns) indexed by strings of length* $l$ *which start and end at*

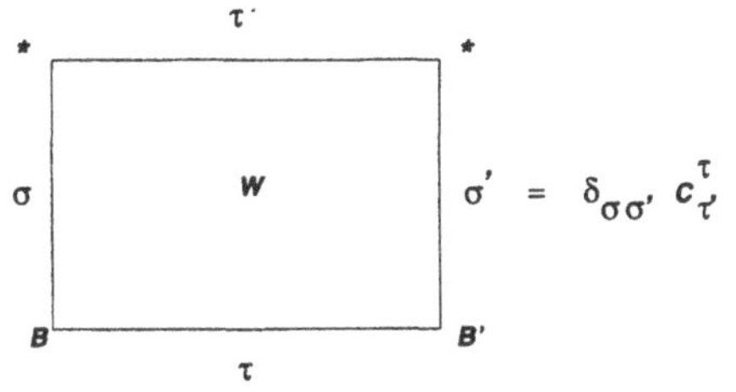


Figure 3.5: Two * flatness

*$B$ (resp., $*$) and which are 'consistent with right-multiplication by $\mathbb{Y}$'.*

*Then $C$ is an isometric matrix - i.e., $C^*C = id$ - which is independent of the string $\sigma$ from $*$ to $B$, and, in particular, also of $k$ and $X_1, \cdots, X_k$.*

**Proof:** (a) Begin by noting that if $\mu : * \otimes * \rightarrow *$ is given by $\mu(a \otimes b) = ab \; \forall a, b \in N$, then $\mu$ is an isomorphism of bimodules.

Let $\sigma\tau \in \mathcal{B}_\gamma$. Thus $\sigma$ (resp., $\tau$) is a string from $*$ to $*$ of length $k$ (resp., $l$) which is 'consistent with left-tensoring by $\mathbb{X}$' (resp., 'consistent with right-tensoring by $\mathbb{Y}$'). Let $L_\sigma$ (resp., $R_\tau$) be as in equation (3.5.18) (resp., equation (3.5.19)). Let us also write $X = X_k \otimes \cdots \otimes X_1$ and $Y = Y_1 \otimes \cdots \otimes Y_l$; thus, for instance, we have:

$$L_\sigma \in Hom(X \otimes *, *) \; , \; R_\tau \in Hom(* \otimes Y, *) \; .$$

Note now that the definitions imply the commutativity of the following diagram:

$$\begin{array}{ccccc} X \otimes * \otimes * \otimes Y & \stackrel{id_X \otimes \mu \otimes id_Y}{\rightarrow} & X \otimes * \otimes Y & & \\ & & L_\sigma \otimes id_Y \downarrow & \searrow id & \\ L_\sigma \otimes R_\tau \downarrow & & * \otimes Y & & X \otimes * \otimes Y \\ & & \downarrow R_\tau & \swarrow (\sigma\tau) & \\ * \otimes * & \stackrel{\mu}{\rightarrow} & * & & \end{array}$$

On the other hand, the symbol $\sigma\tau$ also makes sense as an element of $\mathcal{B}_{\gamma'}$; in order to avoid confusion, let us write $(\sigma\tau)'$ for this element. Thus, we have $(\sigma\tau)' = L_\sigma \circ (id_X \otimes R_\tau)$, while $(\sigma\tau) = R_\tau \circ (L_\sigma \otimes id_Y)$.

Exactly as before, we also have the following commutative diagram:

$$\begin{array}{ccccc}
X \otimes * \otimes * \otimes Y & \stackrel{id_X \otimes \mu \otimes id_Y}{\rightarrow} & X \otimes * \otimes Y & & \\
 & & id_X \otimes R_\tau \downarrow & \searrow id & \\
L_\sigma \otimes R_\tau \downarrow & & X \otimes * & & X \otimes * \otimes Y \\
 & & \downarrow L_\sigma & \swarrow (\sigma\tau)' & \\
* \otimes * & \stackrel{\mu}{\rightarrow} & * & &
\end{array}$$

Comparing these two diagrams, we see that $\sigma\tau = (\sigma\tau)'$, and hence,

$$W^{\sigma\tau}_{\sigma'\tau'} = \langle \sigma\tau, \sigma'\tau' \rangle = \langle (\sigma\tau)', \sigma'\tau' \rangle = \delta_{\sigma,\sigma'}\delta_{\tau,\tau'} ,$$

thereby establishing (a).

(b) Fix $B' \in \mathcal{M}$, and consider the following four marked courses which start at $*$ and finish at $B'$:

$$\begin{array}{rcl}
\gamma = ((k,0),(0,l),\mathbb{X},\mathbb{Y}) & , & \gamma' = ((k),(l),\mathbb{X},\mathbb{Y}) \\
\gamma_1 = ((k_1,0),(0,l),\mathbb{X}^{(1)},\mathbb{Y}) & , & \gamma_1' = ((k_1),(l),\mathbb{X}^{(1)},\mathbb{Y}) .
\end{array}$$

Fix some choice of bases $\mathcal{B}_\gamma, \mathcal{B}_{\gamma'}, \mathcal{B}_{\gamma_1'}, \mathcal{B}_{\gamma_1'}$. Fix $B \in \mathcal{M}$ and consider the typical element of $\mathcal{B}_\gamma$ (resp., $\mathcal{B}_{\gamma_1}$) whose profile $\mathbb{A} = (* = A_0, \cdots, A_{k+l} = B')$ (resp., $\mathbb{A}^{(1)} = (A_0^{(1)} = *, \cdots, A_{k_1+l}^{(1)} = B')$) satisfies $A_k = B$ (resp., $A_{k_1} = B$); let us call this $\sigma\tau$ (resp., $\sigma^{(1)}\tau$). Similarly, let us write $\sigma'\tau'$ and $\sigma^{(1)\prime}\tau^{(1)\prime}$ for the typical element of $\mathcal{B}_{\gamma'}$ (resp., $\mathcal{B}_{\gamma_1'}$) whose profile $\mathbb{A}' = (* = A_0', \cdots, A_{k+l}' = B')$ (resp., $\mathbb{A}^{(1)\prime} = (A_0^{(1)\prime} = *, \cdots, A_{k_1+l}^{(1)\prime} = B')$) satisfies $A_l = *$.

Now consider the marked tethered courses $\Gamma = ((k+k_1, 0), (0,l), \mathbb{X} \circ \widetilde{\mathbb{X}^{(1)}}, \mathbb{Y})$ and $\Gamma' = ((k+k_1),(l),\mathbb{X} \circ \widetilde{\mathbb{X}^{(1)}}, \mathbb{Y})$ (where we write $\mathbb{X} \circ \widetilde{\mathbb{X}^{(1)}} = (X_1, \cdots, X_k, \overline{X_1^{(1)}}, \cdots, \overline{X_{k_1}^{(1)}})$) both of which start and finish at $*$; and note that (we may choose bases for the intertwiner spaces for $\Gamma, \Gamma'$ in such a way that) $(\sigma \circ \sigma_\updownarrow^{(1)})\tau^{(1)\prime} \in \mathcal{B}_\Gamma$, $(\sigma' \circ \sigma_\updownarrow^{(1)\prime})\tau' \in \mathcal{B}_\Gamma'$ (where the symbol $(\sigma \circ \sigma_\updownarrow^{(1)})$ denotes $(\sigma_1, \cdots, \sigma_k, \sigma_{k_1\updownarrow}^{(1)}, \cdots, \sigma_{1\updownarrow}^{(1)})$.

Deduce from the already proved part (a) of this proposition, Remark 3.5.2(d), and the second equation in Remark 3.5.3(b) that

$$\begin{aligned}\delta_{\sigma,\sigma'}\delta_{\sigma^{(1)},\sigma^{(1)'}}\delta_{\tau^{(1)'},\tau'} &= W^{(\sigma\circ\sigma^{(1)}_{\ddagger})\tau^{(1)'}}_{(\sigma'\circ\sigma^{(1)'}_{\ddagger})\tau'} \\ &= \sum_{\tau} W^{\sigma\tau}_{\sigma'\tau'} W^{\sigma^{(1)}_{\ddagger}\tau^{(1)'}}_{\sigma^{(1)'}_{\ddagger}\tau} \\ &= \sum_{\tau} W^{\sigma\tau}_{\sigma'\tau'} \sqrt{\frac{dim_N(B')}{dim_N(B)}}\ \overline{W^{\sigma^{(1)}\tau}_{\sigma^{(1)'}\tau^{(1)'}}}\ ,\end{aligned} \tag{3.5.23}$$

where the sum in the last two lines above range over all strings $\tau$ of length $l$ from $B$ to $B'$ which are 'consistent with right-tensoring by $\mathbb{Y}$'.

Set $\sigma^{(1)} = \sigma,\ \sigma^{(1)'} = \sigma',\ \tau^{(1)'} = \tau'$ in equation (3.5.23) to first deduce that if $\sigma \neq \sigma'$, then

$$0 = \sum_{\tau''} \sqrt{\frac{dim_N(B')}{dim_N(B)}}\ |W^{\sigma\tau}_{\sigma'\tau'}|^2\ ,$$

and in particular, $W^{\sigma\tau}_{\sigma'\tau'} = 0$, thereby establishing the first half of the statement (i) of (b).

Next, for any fixed $\sigma$ as above, let us write $C(\sigma)$ for the matrix given by $C(\sigma) = ((C(\sigma)^{\tau}_{\tau'}))$. Note then that if we set $\sigma' = \sigma, \sigma^{(1)'} = \sigma^{(1)}$ in equation (3.5.23), then ($B = B'$ so that the square root term disappears and) we may re-write that equation as

$$\begin{aligned}\delta_{\tau^{(1)'},\tau'} &= \sum_{\tau} C(\sigma)^{\tau}_{\tau'}\overline{C(\sigma^{(1)})^{\tau}_{\tau^{(1)'}}} \\ &= (\ C(\sigma^{(1)})^{*}C(\sigma)\ )^{\tau^{(1)'}}_{\tau'}\ .\end{aligned}$$

The above equation says simultaneously that $C(\sigma)$ is an isometric matrix, and that $C(\sigma) = C(\sigma^{(1)})$. (*Reason*: Since $C(\sigma^{(1)})$ is also an isometry, it follows from the preceding equation that $C(\sigma^{(1)})^*$ agrees with $C(\sigma)^*$ on its initial space, and by symmetry also $C(\sigma)^*$ agrees with $C(\sigma^{(1)})^*$ on its initial space.) The proof of the proposition is complete.

□

REMARK 3.5.5 (a) There is a dual version of (b) and (c) above, where it is required that $s(\tau) = s(\tau') = *$; in this case, we have

$$W^{\sigma\tau}_{\sigma'\tau'} = \delta_{\tau,\tau'} D^{\sigma}_{\sigma'} ,$$

where the matrix $D = ((D^{\sigma}_{\sigma'}))$ is an isometric matrix which is independent of $\tau$. The proof is essentially the same as the one given above.

(b) There is another generalisation of the flatness proposition as follows. Suppose now that we allow bimodules to be chosen from $\cup_{i,j=0}^{1} \mathcal{M}_{ij}$ (rather than just $\mathcal{M}_{00}$). In this case, one defines the $W$-matrix - for $k = l = 1$ - exactly as in equation (3.4.11), but the square root terms occurring in the definition of the $Z$-function need to be handled a little more carefully. (Of course, it is understood that now, if we have the $1 \times 1$ rectangle

$$\begin{array}{ccccc} & & \cdot Y & & \\ & A & \rightarrow & C & \\ X\cdot & \downarrow & & \downarrow & X\cdot \\ & B & \rightarrow & D & \\ & & \cdot Y & & \end{array} ,$$

it is assumed that there exist $i, j, k, l \in \{0, 1\}$ such that $X \in \mathcal{M}_{ij}, A \in \mathcal{M}_{jk}, Y \in \mathcal{M}_{kl}, D \in \mathcal{M}_{il}, B \in \mathcal{M}_{ik}, C \in \mathcal{M}_{jl}$; only then will all the relevant tensor-products make sense.)

Then we set

$$|V| = \sqrt{(dim\ {}_{M_i}V)(dim\ V_{M_j})}$$

for $V \in \mathcal{M}_{ij}$ - where we write $M_i$ to denote $N$ or $M$ according as $i = 0$ or $i = 1$ - and if the $1 \times 1$ rectangle above is labelled with intertwiners $\xi_1 \in \mathcal{H}^{B}_{XA}, \xi_2 \in \mathcal{H}^{D}_{XC}, \xi_3 \in \mathcal{H}^{D}_{BY}, \xi_4 \in \mathcal{H}^{C}_{AY}$, we define

$$Z(\xi_1, \xi_2, \xi_3, \xi_4) = \sqrt{\frac{|B|\ |C|}{|A|\ |D|}}\, W(\xi_1, \xi_2, \xi_3, \xi_4) .$$

It is then true that this also defines a 'quantum 6j symbol' - the reason for the 'quotes' being that what we now have is not quite a fusion algebra but something like an '$M_2$-graded fusion algebra' and bimodules have left- and right-attributes which must match before their products can be formed. What we mean is that we have the appropriate version of an $S_4$-action and appropriate invariance of the $Z$-function defined as above.

And the flatness proposition is valid essentially as stated, only now each of (a), (b) and (c) stand for two statements, one each with (all occurrences of) $*$ being replaced by $*_N = {}_N L^2(N)_N$ and $*_N = {}_M L^2(M)_M$, and the bimodules (such as $X_i, Y_j, s(\sigma_i), s(\tau_j)$) that feature in the big rectangles can all be any of the four kinds of bimodules subject only to the 'matching-attribute requirement for neighbours' (for instance, if $X_i \in \mathcal{M}_{jk}$, then we must have $s(\sigma_i) \in \mathcal{M}_{kl}$ for some $l$, so that $X_i \otimes_{M_k} s(\sigma_i)$ makes sense).

Only the following lemma stands between us and the fact that the $Z$-function, as defined by equation (3.4.11), satisfies (the (2-3 condition) and hence satisfies) all the requirements of a quantum 6j symbol.

LEMMA 3.5.6 *With the foregoing notation, we have*

$$\sum_{\xi_3} Z(\xi_1, \xi_2, \xi_3, \xi_4)\ Z(\eta_1, \xi_3, \eta_3, \eta_4) \ =$$

$$\sum_{G,\lambda,\mu,\nu} d(G) Z(\lambda, \xi_4, \mu, \eta_4) Z(\xi_1, \nu, \eta_1, \lambda) Z(\nu, \xi_2, \eta_3, \mu) \ ;$$

*for arbitrary* $A, B, C, D, E, F, X, Y \in \mathcal{M}$ *and* $\xi_1 \in \mathcal{H}^B_{XA}, \xi_2 \in \mathcal{H}^D_{XC}, \xi_3 \in \mathcal{H}^D_{BY}, \xi_4 \in \mathcal{H}^C_{AY}, \eta_1 \in \mathcal{H}^H_{BE}, \eta_3 \in \mathcal{H}^D_{HF}, \eta_4 \in \mathcal{H}^Y_{EF}$, *where the sum on the right ranges over* $G \in \mathcal{M}$, $\lambda \in \mathcal{B}_5, \mu \in \mathcal{B}_6, \nu \in \mathcal{B}_7$ *where* $\mathcal{B}_5, \mathcal{B}_6$ *and* $\mathcal{B}_7$ *are orthonormal bases for* $\mathcal{H}^G_{AE}, \mathcal{H}^C_{GF}, \mathcal{H}^Y_{EF}$ *respectively.*

**Proof :** The fact is that this identity is more or less a re-formulation of the flatness condition expressed in Proposition 3.5.4(a) in the case when $k = l = 3$.

We first need some notation. To start with, we shall, for typographical convenience, simply write $|V|$ for $dim_N(V)$ for all $V \in \mathcal{M}(= \mathcal{M}_{00})$. Next, for any $V \in \mathcal{M}$, note that there exist unique isomorphisms

$${}_V 1 : V \otimes * \rightarrow V \text{ and } 1_V : * \otimes V \rightarrow V$$

satisfying

$$ {}_V1(\eta \otimes y) = \eta \cdot y \text{ and } 1_V(x \otimes \eta) = x \cdot \eta \ \forall x, y \in N, \eta \in V^0 \ . $$

Before proceeding further, we need to make some observations regarding the above intertwiners. Notice first that for arbitrary $X, Y \in \mathcal{M}$, we have $({}_X1 \otimes id_Y) \circ (id_X \otimes 1_Y^*) = id_{X \otimes Y}$; hence, if we have a rectangle of the form

$$ \begin{array}{ccccccc} & & & \cdot Y & & & \\ & & * & \stackrel{1_Y^*}{\rightarrow} & Y & & \\ X\cdot & {}_X1 \downarrow & & & \downarrow \sigma & X\cdot & , \\ & & X & \stackrel{\tau}{\rightarrow} & D & & \\ & & & \cdot Y & & & \end{array} $$

where $\sigma$ and $\tau$ belong to some orthonormal basis for $\mathcal{H}_{XY}^D$, then the associated $W$-value is given by

$$ \begin{aligned} W_{\sigma 1_Y}^{1_X \tau} &= \tau \circ ({}_X1 \otimes id_Y) \circ (id_X \otimes 1_Y^*) \circ \sigma^* \\ &= \tau \sigma^* \\ &= \delta_{\sigma\tau} \ . \end{aligned} $$

Then in view of the equations displayed in Remark 3.5.3(b), we have the following identities, valid for arbitrary $\sigma, \tau \in \mathcal{B}_{XY}^D$:

$$ \begin{aligned} W_{\sigma 1_Y}^{{}_X1 \tau} &= \sqrt{\frac{|X|\ |Y|}{|D|}}\ \overline{W}_{{}_X1 1_{Y\leftrightarrow}}^{\sigma \tau_{\leftrightarrow}} \\ &= W_{{}_X1_{\updownarrow} \tau_{Y\leftrightarrow}}^{\sigma_{\updownarrow} 1_{Y\leftrightarrow}} && (3.5.24) \\ &= \sqrt{\frac{|X|\ |Y|}{|D|}}\ \overline{W}_{\sigma_{\updownarrow} \tau}^{{}_X1_{\updownarrow} 1_Y} \\ &= \delta_{\sigma,\tau} \ . && (3.5.25) \end{aligned} $$

Now, consider the marked tethered courses of size $3 \times 3$ given by $\gamma = ((3,0), (0,3), (A, X, \bar{B}), (Y, \bar{F}, \bar{E}))$ and $\gamma' = ((3), (3), (A, X, \bar{B}), (Y, \bar{F}, \bar{E}))$, which both start and finish at $*$.

Assume - as we may - that our bases $\mathcal{B}_{XY}^Z$'s have been so chosen that $\sigma\tau \in \mathcal{B}_\gamma, \sigma'\tau' \in \mathcal{B}_{\gamma'}$, where $\sigma = ({}_A1, \xi_1, 1_{B\updownarrow}) = \sigma'$, $\tau = (1_Y, \eta_{4\leftrightarrow}, 1_{E\leftrightarrow}) = \tau'$. Notice that since $\sigma = \sigma', \tau = \tau'$,

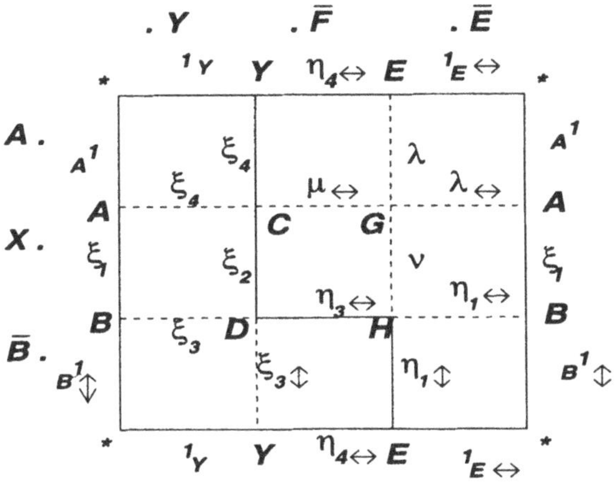


Figure 3.6: 3 x 3 flatness

the element $(\sigma'\tau') \in \mathcal{B}_{\gamma'}$ is exactly what we called $(\sigma\tau)'$ in the proof of Proposition 3.5.4(a).

Consider the 'intermediate' marked tethered course given by $\gamma^{(1)} = ((2,1,0),(1,1,1),(A,X,\bar{B}),(Y,\bar{F},\bar{E}))$ (with source and range equal to $*$). Note that the typical element $\sigma^{(1)}\tau^{(1)} \in \mathcal{B}_{\gamma^{(1)}}$ is given by $\sigma^{(1)} = (\xi_4, \xi_2, \eta_{1\updownarrow})$ and $\tau^{(1)} = (1_Y, \eta_{3\leftrightarrow}, 1_{E\leftrightarrow})$.

In view of Remark 3.5.2(b), we have

$$W^{\gamma}_{\gamma'} = W^{\gamma}_{\gamma^{(1)}} W^{\gamma^{(1)}}_{\gamma'}$$

so that

$$W^{\sigma\tau}_{(\sigma\tau)'} = \sum_{\sigma^{(1)}\tau^{(1)}} W^{\sigma\tau}_{\sigma^{(1)}\tau^{(1)}} W^{\sigma^{(1)}\tau^{(1)}}_{(\sigma\tau)'} .$$

On the one hand, (with $(\sigma^{(1)}\tau^{(1)})$ as above), in view of Remark 3.5.3(d) and equation (3.5.24), we have

$$\begin{aligned} W^{\sigma\tau}_{\sigma^{(1)}\tau^{(1)}} &= \sum_{\psi_1,\psi_2,\psi_3} W^{A 1 \psi_1}_{\xi_4 1_Y} W^{\xi_1\psi_2}_{\xi_2\psi_1} W^{B 1_{\updownarrow} 1_Y}_{\psi_3\psi_2} W^{\psi_3\eta_{4\leftrightarrow}}_{\eta_{1\updownarrow}\eta_{3\leftrightarrow}} \\ &= \sum_{\psi_1,\psi_2,\psi_3} \delta_{\psi_1,\xi_4} W^{\xi_1\psi_2}_{\xi_2\psi_1} \sqrt{\frac{|D|}{|B|\,|Y|}} \delta_{\psi_3,\psi_{2\updownarrow}} W^{\psi_3\eta_{4\leftrightarrow}}_{\eta_{1\updownarrow}\eta_{3\leftrightarrow}} \\ &= \sqrt{\frac{|D|}{|B|\,|Y|}} \sum_{\xi_3} W^{\xi_1\xi_3}_{\xi_2\xi_4} W^{\xi_{3\updownarrow}\eta_{4\leftrightarrow}}_{\eta_{1\updownarrow}\eta_{3\leftrightarrow}} . \qquad (3.5.26) \end{aligned}$$

Similar reasoning yields

$$
\begin{aligned}
& W^{\sigma^{(1)}\tau^{(1)}}_{(\sigma\tau)'} \\
& = \sum_{G,\phi_1,\phi_2,\phi_3,\phi_4,\phi_5} W^{\xi_4\phi_2}_{\phi_1\eta_{4\leftrightarrow}} W^{\xi_2\eta_{3\leftrightarrow}}_{\phi_3\phi_2} W^{\phi_1\phi_4}_{A^{1}1_{E\leftrightarrow}} W^{\phi_3\phi_5}_{\xi_1\phi_4} W^{\eta_{1\updownarrow}1_{E\leftrightarrow}}_{B^{1}{}_{\updownarrow}\phi_5} \\
& = \sum_{G,\phi_1,\phi_2,\phi_3,\phi_4,\phi_5} W^{\xi_4\phi_2}_{\phi_1\eta_{4\leftrightarrow}} W^{\xi_2\eta_{3\leftrightarrow}}_{\phi_3\phi_2} \delta_{\phi_1,\phi_{4\leftrightarrow}} \sqrt{\frac{|G|}{|E|\ |A|}} W^{\phi_3\phi_5}_{\xi_1\phi_4} \delta_{\eta_1,\phi_{5\leftrightarrow}} \\
& = \sum_{G,\lambda,\mu,\nu} \sqrt{\frac{|G|}{|E|\ |A|}} W^{\xi_4\mu_{\leftrightarrow}}_{\lambda\eta_{4\leftrightarrow}} W^{\xi_2\eta_{3\leftrightarrow}}_{\nu\mu_{\leftrightarrow}} W^{\nu\eta_{1\leftrightarrow}}_{\xi_1\lambda_{\leftrightarrow}} . \qquad (3.5.27)
\end{aligned}
$$

The equations displayed in Remark 3.5.3(d) imply that

$$
\begin{aligned}
W^{\xi_4\mu_{\leftrightarrow}}_{\lambda\eta_{4\leftrightarrow}} &= \sqrt{\frac{|C|\ |E|}{|Y|\ |G|}} \overline{W^{\lambda\mu}_{\xi_4\eta_4}} \\
W^{\xi_2\eta_{3\leftrightarrow}}_{\nu\mu_{\leftrightarrow}} &= \sqrt{\frac{|D|\ |G|}{|C|\ |H|}} \overline{W^{\nu\eta_3}_{\xi_2\mu}} \\
W^{\nu\eta_{1\leftrightarrow}}_{\xi_1\lambda_{\leftrightarrow}} &= \sqrt{\frac{|H|\ |A|}{|G|\ |B|}} \overline{W^{\xi_1\eta_1}_{\nu\lambda}} \\
W^{\xi_{3\updownarrow}\eta_{4\leftrightarrow}}_{\eta_{1\updownarrow}\eta_{3\leftrightarrow}} &= W^{\eta_1\eta_3}_{\xi_3\eta_4} ;
\end{aligned}
$$

and hence, we may re-write equations (3.5.26) and (3.5.27) as:

$$
\begin{aligned}
& W^{\sigma\tau}_{\sigma^{(1)}\tau^{(1)}} \\
& = \sqrt{\frac{|D|}{|B|\ |Y|}} \sum_{\xi_3} W^{\xi_1\xi_3}_{\xi_2\xi_4} W^{\eta_1\eta_3}_{\xi_3\eta_4} \\
& = \sqrt{\frac{|D|}{|B|\ |Y|}} \sum_{\xi_3} \sqrt{|B|\ |C|} Z(\xi_1,\xi_2,\xi_3,\xi_4) \ \times \\
& \qquad \times \ \sqrt{|H|\ |Y|} Z(\eta_1,\xi_3,\eta_3,\eta_4) \\
& = \sqrt{|C|\ |D|\ |H|} \sum_{\xi_3} Z(\xi_1,\xi_2,\xi_3,\xi_4) Z(\eta_1,\xi_3,\eta_3,\eta_4)
\end{aligned}
$$

and

$$
\begin{aligned}
W_{(\sigma\tau)'}^{\sigma^{(1)}\tau^{(1)}} & \\
&= \sum_{G,\lambda,\mu,\nu} \sqrt{\frac{|G|}{|E|\,|A|}}\sqrt{\frac{|C|\,|E|}{|Y|\,|G|}}\sqrt{\frac{|D|\,|G|}{|C|\,|H|}} \times \\
&\qquad \times \sqrt{\frac{|H|\,|A|}{|G|\,|B|}}\; \overline{W_{\xi_4\eta_4}^{\lambda\mu} W_{\xi_2\mu}^{\nu\eta_3} W_{\nu\lambda}^{\xi_1\eta_1}} \\
&= \sqrt{\frac{|D|}{|B|\,|Y|}} \sum_{G,\lambda,\mu,\nu} \overline{W_{\xi_4\eta_4}^{\lambda\mu} W_{\xi_2\mu}^{\nu\eta_3} W_{\nu\lambda}^{\xi_1\eta_1}} \\
&= \sqrt{|C|\,|D|\,|H|} \sum_{G,\lambda,\mu,\nu} |G|\, \overline{Z(\lambda,\xi_4,\mu,\eta_4)} \\
&\qquad \times \overline{Z(\nu,\xi_2,\eta_3,\mu)}\; \overline{Z(\xi_1,\nu,\eta_1,\lambda)}\,.
\end{aligned}
$$

Thus, the identity in the (2-3 condition) is seen to translate, in our notation, to the statement that

$$
W_{\sigma^{(1)}\tau^{(1)}}^{\sigma\tau} = \overline{W_{(\sigma\tau)'}^{\sigma^{(1)}\tau^{(1)}}} \tag{3.5.28}
$$

for all possible $(\sigma\tau) \in \mathcal{B}_\gamma, \sigma^{(1)}\tau^{(1)} \in \mathcal{B}_{\gamma^{(1)}}$; but $3 \times 3$ flatness implies that

$$
\begin{aligned}
1 &= W_{(\sigma\tau)'}^{\sigma\tau} \\
&= \sum_{\sigma^{(1)}\tau^{(1)}} W_{\sigma^{(1)}\tau^{(1)}}^{\sigma\tau} W_{(\sigma\tau)'}^{\sigma^{(1)}\tau^{(1)}} \;;
\end{aligned}
$$

and the validity of equation (3.5.28) is a consequence now of the fact that two unit vectors can have inner-product equal to one if and only if they are equal; and the lemma is proved.

□

# Chapter 4

# The group example

## 4.1 The TQFT

Throughout this section, we shall fix a finite group $G$, the identity of which will be denoted simply by 1.

Given an 'outer action' of $G$ on a $II_1$ factor $P$, there are two naturally arising subfactors, namely, the inclusions given by $N = P \subset P \rtimes G = M$ and $N = P^G \subset P = M$. It turns out - see [EK] - that both these subfactors give rise to the same TQFT. We shall endeavour to work out various details in this section concerning the TQFT coming from the first kind of subfactor above.

The first objects to spell out are the underlying fusion algebra and the associated system of intertwiner spaces. The fusion algebra is just the complex group algebra $\mathcal{A} = \mathbb{C}G$ with distinguished basis $\mathcal{B} = G$ as in Example 1.3.5(a). Recall that the 'dimension function' here is the trivial function defined by $d_g = 1 \; \forall \; g \in G$ - so that $w(\mathcal{A}) = |G|$ - and that the structure constants are given by

$$N_{gh}^k = \begin{cases} 1 & \text{if } gh = k \\ 0 & \text{otherwise} \end{cases} .$$

The 'system of intertwiner spaces' is equally trivial, and defined by

$$\mathcal{H}_{gh}^k = \begin{cases} \mathbb{C} & \text{if } gh = k \\ \{0\} & \text{otherwise} \end{cases} .$$

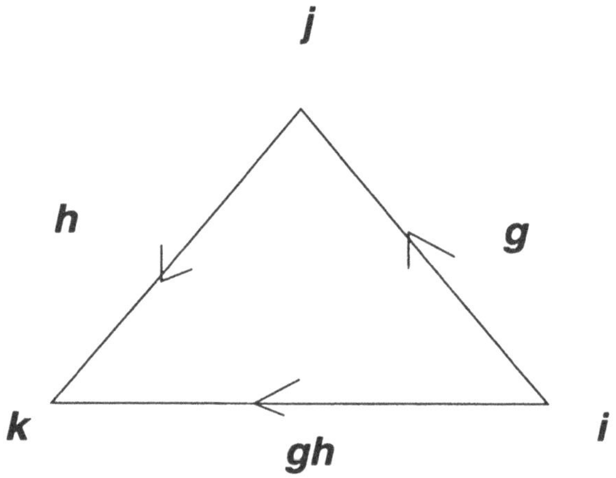


Figure 4.1: A compatible state

We shall assume that when the Hilbert space $\mathcal{H}^k_{gh}$ is non-zero, the basis is chosen to be $\mathcal{B}^k_{gh} = \{1\}$, and that the Frobenius reciprocity map $F_\sigma$ is the identity or the complex-conjugation map according as $\sigma$ is even or odd.

Finally, we define the quantum $6j$ symbol by demanding that $Z(\phi)$ is always zero or one for any state $\phi$, and that the value is 1 precisely when the state $\phi$ of the tetrahedron $T$ is *compatible with* $G$ in the following sense: if $f = (ijk)$ is a face with $i <_\phi j <_\phi k$ , if $\phi^1(ij) = g$ and $\phi^1(jk) = h$, then $\phi^1(ik) = gh$.

More generally, if $M$ is any compact manifold (possibly with boundary), if $\Delta$ is a triangulation of $M$, and if $\leq$ is some total order specified on the set of vertices of $\Delta$, we shall think of the edge-set $E(\Delta)$, with every edge being directed from the 'smaller' vertex to the 'larger'; and we shall use the word *colouring* to denote a function $\phi$ from the set of edges of $\Delta$ to $G$ which is 'compatible with $G$' exactly as above, in the sense that if $f = (ijk)$ is any triangle in $\Delta$, with $i <_\phi j <_\phi k$ , and if $\phi^1(ij) = g$ and $\phi^1(jk) = h$, then $\phi^1(ik) = gh$. We shall use the notation $Col(\Delta)$ to signify the set of colourings of $(\Delta, \leq)$.

It is not too hard to see that if $\phi \in Col(\Delta)$, there exists a unique (groupoid-) homomorphism* $\tilde{\phi} : EPG(\Delta) \to G$ such

*See §A.2 for the edge-path groupoid $EPG(\Delta)$.

that if $e = (u, v)$ is any edge in $\Delta$, then $\tilde{\phi}(e) = \phi(e)^{\epsilon}$ where $\epsilon = +1$ or $\epsilon = -1$ according to whether $u \leq v$ or $v \leq u$; further, it is seen that the correspondence $\phi \leftrightarrow \tilde{\phi}$ establishes a bijection between $Col(\Delta)$ and $Hom(EPG(\Delta), G)$.

Before proceeding to the statement of the next lemma, observe that the fundamental group of a manifold (relative to any base-point) is a subset (in fact a sub-groupoid) of its fundamental groupoid, which in turn may be identified - see §A.2 - with the edge-path groupoid of any of its triangulations; thus, for an arbitrary vertex $v$ of $\Delta$, we have a natural inclusion map (in fact a groupoid map) $j_v : \pi_1(M, v) \to EPG(\Delta)$, which induces, by restriction, the map $(j_v)^* : Hom(EPG(\Delta), G) \to Hom(\pi_1(M, v), G)$. (The symbol $\pi_1(M, v)$ denotes, naturally, the 'fundamental group of $M$ with base point $v$'.)

LEMMA 4.1.1 *(a) Any spanning tree $\Gamma$ in the graph $E(\Delta)$ induces a bijection*

$$F_\Gamma : EPG(\Delta) \to (Vert(\Delta) \times \pi_1(M, v) \times Vert(\Delta)) ,$$

*which is given by*

$$F_\Gamma([\gamma]) = (v_0, [\gamma_0] \cdot [\gamma] \cdot [\gamma_1]^{-1}, v_1) , \tag{4.1.1}$$

*where $s(\gamma) = v_0$, $r(\gamma) = v_1$, and $\gamma_i$, $i = 0, 1$ is the unique shortest path from $v$ to $v_i$ which lies entirely in $\Gamma$.*

*(b)*

$$\begin{aligned} |Col(\Delta)| &= |Hom(EPG(\Delta), G)| \\ &= |G|^{Vert(\Delta)-1} \, |Hom(\pi_1(M), G)| \, . \end{aligned}$$

**Proof:** (a) This is a straightforward consequence of the fact that any two edge-paths which lie entirely in a tree are homotopic rel. end-points if and only if they have the same end-points.

(b) Let $Col(\Delta) \ni \phi \leftrightarrow \tilde{\phi} \in Hom(EPG(\Delta), G)$ as in the discussion preceding this lemma. It should be clear from (a) above that the mapping defined on $Col(\Delta)$ by

$$\phi \mapsto (\phi|_{E(\Gamma)} , \tilde{\phi}|_{\pi_1(M,v)}) \in G^{E(\Gamma)} \times Hom(\pi_1(M, v), G) \tag{4.1.2}$$

is a bijection - where $E(\Gamma)$ denotes the set of $(|Vert(\Delta)| - 1)$ edges of $\Gamma$ - and the desired conclusion follows. □

PROPOSITION 4.1.2 *If $M$ is a connected closed oriented 3-manifold, then*

$$\langle M \rangle = \frac{|Hom(\pi_1(M), G)|}{|G|} .$$

**Proof:** If $(\Delta, \leq)$ is a choice of triangulation and total order on a 3-manifold $M$, the triviality of the quantum $6j$ symbol and the definitions are seen to imply that

$$\langle M \rangle = \frac{|Col(\Delta)|}{|G|^{|Vert(\Delta)|}} , \tag{4.1.3}$$

where $Col(\Delta)$ denotes the set of all 'colourings of $\Delta$'. (This uses the fact that the dimension function is also trivial in this case ($d(g) = 1 \ \forall \ g \in G$) - see Example 1.3.5(a).)

In view of Lemma 4.1.1(b), it is clear that the desired equality is a consequence of equation 4.1.3. □

We turn next to the Hilbert space associated to a closed oriented surface $\Sigma$; we assume that $\Sigma$ is connected and that $(\delta, \leq)$ is some choice of a triangulation $\delta$ and total order $\leq$ on $\Sigma$. In the notation of equation 2.2.8, it is clear from the 'triviality' of this example (and its consequence that the intertwiner spaces have dimension 0 or 1) that the Hilbert space $W(\Sigma, \delta, \leq)$ has an orthonormal basis - call it $\{\xi_\phi : \phi \in Col(\delta)\}$ - indexed by the collection $Col(\delta)$ of colourings on $\delta$. Consequently, the projection operator $p_{(\Sigma,\delta)}$ (see Lemma 2.2.9) is represented by a matrix $p = ((p^{\phi_0}_{\phi_1}))$ with rows and columns indexed by $Col(\delta)$.

In the sequel, colourings, i.e., elements of $Col(\delta)$, will be denoted by $\phi$ (suitably adorned with subscripts or in capitals), and the associated homomorphism $\tilde{\phi}|_{\pi_1(\Sigma,v)} \in Hom(\pi_1(\Sigma, v), G)$ will be denoted by (appropriate versions of) $\psi$. Notice that there is a natural action of $G$ on $Hom(\pi_1(\Sigma), G)$, given by $(g \cdot \psi)([\gamma]) = g\psi([\gamma])g^{-1}$; we shall consider two colourings - say $\phi_i, i = 0, 1$ - as being *equivalent* if the associated homomorphisms $\psi_i = \tilde{\phi}_i|_{\pi_1(\Sigma,v)}$ are in the same $G$-orbit, and use (at least in the following lemma) the notation $[\phi]$ to denote the equivalence class of the colouring $\phi$.

PROPOSITION 4.1.3 *With the foregoing notation, we have:*

$$p_{\phi_1}^{\phi_0} = \begin{cases} [\phi_0]^{-1} & \text{if } [\phi_0] = [\phi_1] \\ 0 & \text{if } [\phi_0] \neq [\phi_1] \end{cases} .$$

*In particular, we have*

$$dim(V(\Sigma)) = \mid Hom(\pi_1(\Sigma), G) \, / \sim \mid . \tag{4.1.4}$$

**Proof:** Fix a vertex $v$ of $\delta$. If $\psi \in Hom(\pi_1(\Sigma, v), G)$, let us write $H_{v,\psi}$ to denote the associated 'isotropy' subgroup of $G$; thus

$$H_{v,\psi} = \{g \in G : g\psi([\gamma])g^{-1} = \psi([\gamma]) \ \forall [\gamma] \in \pi_1(\Sigma)\} .$$

It then follows from Lemma 4.1.1 that

$$|[\phi]| = |G|^{Vert(\delta)-1} \, [G : H_{v,\psi}] = |G|^{Vert(\delta)} \, |H_{v,\psi}|^{-1} , \tag{4.1.5}$$

where we write $\psi = \tilde{\phi}|_{\pi_1(\Sigma,v)}$.

Let us write $M = \Sigma \times [0,1]$. For a (chosen and fixed) $(\delta, \leq)$ as above, let $(\Delta, <<)$ be chosen subject only to the following restrictions:

(a) $\Delta$ is some* triangulation of $M$ such that: (i) $\Delta|_{\Sigma \times \{i\}} = \delta \times \{i\}$, for $i = 0, 1$; (ii) $\Delta$ contains no vertices in $\Sigma \times (0,1)$; and (iii) the line-segment $\{u\} \times [0,1]$ is a 1-simplex in $\Delta$, for every 0-simplex $u$ in $\delta$; and

(b) $<<$ is the total order on $Vert(\Delta) = Vert(\delta) \times \{0,1\}$ defined by

$$(u,i) << (v,j) \Leftrightarrow \begin{cases} u, v \text{ arbitrary} & if \ i < j \\ u \leq v & if \ i = j \end{cases}$$

We see then that (in the notation of Chapter 2)

$$\begin{aligned} p_{\phi_1}^{\phi_0} &= \langle Z_{(\Sigma \times [0,1], \delta \coprod \delta)} \xi_{\phi_0}, \xi_{\phi_1} \rangle \ \text{(see Lemma 2.2.9)} \\ &= \langle (M, \delta \coprod \delta, <<) : \phi_0 \coprod \phi_1 \rangle \ \text{(see Proposition 2.2.6)} \\ &= |G|^{-\ Vert(\delta)} \ |\{\Phi : \Phi \in Col(\Delta), \Phi|_{\delta \times \{i\}} = \phi_i \ , \ i = 0, 1\}| , \end{aligned}$$

*One such triangulation is prescribed by requiring that a three simplex is either contained in $\delta \times \{i\}$ for some $i \in \{0,1\}$, or is given by one of the following alternatives, where $\{u, v, w\}$ is a 2-simplex in $\delta$, with $u \leq v \leq w$: (i) $\{(u,0), (v,0), (w,0), (w,1)\}$; (ii) $\{(u,0), (v,0), (v,1), (w,1)\}$; (iii) $\{(u,0), (u,1), (v,1), (w,1)\}$.

by equation (2.2.11).

In view of equation 4.1.5, we need to check that

$$\begin{aligned} &|\{\Phi : \Phi \in Col(\Delta), \Phi|_{\delta \times \{i\}} = \phi_i \ , \ i = 0,1\}| \\ &= \begin{cases} |H_{v,\psi_0}| & \text{if } [\phi_0] \neq [\phi_1] \\ 0 & \text{otherwise} \end{cases} . \end{aligned} \quad (4.1.6)$$

To do this, we shall appeal to the bijection (4.1.2), applied to the triangulation $\Delta$ of $M$. So fix a spanning tree $\Gamma_0$ of $\delta$, and let $\Gamma$ denote the spanning tree of $\Delta$ defined by

$$\Gamma = (\Gamma_0 \times \{0,1\}) \coprod (\{v\} \times [0,1]) ,$$

where $v$ is a fixed 0-simplex of $\delta$.

We shall find the following maps useful: let $j_t : \Sigma \to M$ be defined by $j_t(x) = (x,t)$.

We appeal to Lemma 4.1.1, applied with $M = \Sigma \times [0,1]$ and $(v,0)$ as the base-point for the manifold $M$; and we use the following fact (which is not too hard to verify (or believe!):

if $\Phi \in Col(\Delta)$ and if $\Phi(e \times \{i\}) = \phi_i(e)$, for every edge $e \in E(\delta)$, then we have:

(i) $\tilde{\Phi}((j_t)_*[\gamma]) = \tilde{\phi_t}([\gamma])$ for all $[\gamma] \in EPG(\Sigma)$, $t = 0,1$; and

(ii) if $\kappa$ is the edge $((v,0),(v,1))$ in $M$, then

$$[\kappa] \ ((j_1)_*[\gamma]) \ [\kappa]^{-1} = (j_0)_*[\gamma] .$$

Applying the homomorphism $\tilde{\Phi}$ to the above equation - and writing $g = \tilde{\Phi}([\kappa])$ - we find that $g \, \tilde{\phi_1}([\gamma]) \, g^{-1} = \tilde{\phi_0}([\gamma])$. Since $\tilde{\phi_t}|_{\pi_1(\Sigma,v)} = \psi_t$, we thus find that if there exists a $\Phi \in Col(\Delta)$ such that $\Phi(e \times \{i\}) = \phi_i(e)$, for every edge $e \in E(\delta)$, then we must have $[\phi_0] = [\phi_1]$.

On the other hand, the maps $(j_t)_* : \pi_1(\Sigma, v) \to \pi_1(M,(v,t))$ are group-isomorphisms, and so the observation (i) above implies that the $\Psi$ component (in the bijection (4.1.2)) of an extending $\Phi$ is uniquely determined (by the corresponding $\psi_t$'s). Finally, the only remaining degree of freedom left for $\Phi|_{E(\Gamma)}$ is what we have called $g$ above, and we have seen that for $\Phi$ to be a colouring, this $g$ must necessarily conjugate $\psi_1$ to $\psi_0$, and the proof is completed by a final appeal to the bijection (4.1.2). □

## 4.2 The quantum double

There is another construction of TQFTs due to Reshetikhin and Turaev (see [Tur]); we will say nothing about this 'surgery method' other than that the input for that construction is something called a *strict modular category.* It turns out that such categories also contain the data of a finite-dimensional fusion algebra, say $\mathcal{A} = \mathbb{C}\mathcal{B}$. One interesting feature of that construction is that the fusion algebra completely determines the dimensions of the Hilbert spaces associated to all closed oriented surfaces; this fact is a consequence of **Verlinde's formula**, which states that for $d = 0, 1, 2, \cdots$, if $\Sigma_d$ denotes the closed oriented surface of genus $d$, then

$$dim\ V_{sur}(\Sigma_d) = \sum_{\pi \in \mathcal{B}} \left( \frac{\mathcal{D}(\mathcal{A})}{d_\pi} \right)^{2d-2} , \tag{4.2.7}$$

where $\mathcal{D}(\mathcal{A})^2$ is what we called the weight (see equation 1.3.20) of $\mathcal{A}$; we have also written $V_{sur}(\cdot)$ for the TQFT obtained by applying the 'surgery method'.

It turns out that any finite group gives rise to a 'strict modular category' via the 'representation theory of the Hopf algebra $D(G)$' given by Drinfeld's quantum double construction. Thus the group $G$ gives rise to (at least) two TQFTs, *viz.*, what we called $V_{sur}$ (when applied to the representation theory of $D(G)$) and the TQFT obtained by the 'triangulation method' as in §4.1, which we shall denote by $V_{tri}$.

We wish to show, in this section, that the dimensions of $V_{tri}(\Sigma_d)$ (as given by equation (4.1.4)) tally with those of $V_{sur}(\Sigma_d)$ (as given by applying equation (4.2.7) to the fusion algebra associated to the representations of $D(G)$).

The *quantum double* of a finite group $G$ is, by definition, the $C^*$-algebra crossed product of the (commutative $C^*$-algebra) $\ell^\infty(G)$ by the conjugation action of $G$; more explicitly, it has a basis given by $\{(x, g) : x, g \in G\}$, and the product and adjoint are given by:

$$\begin{aligned} (x, g) \cdot (y, h) &= \delta_{x, gyg^{-1}}\ (x, gh)\ ;\ \text{and} \\ (x, g)^* &= (g^{-1}xg, g^{-1})\ . \end{aligned}$$

LEMMA 4.2.1 *Let* $f = \sum_{x,a \in G} C(x,a)\ (x,a)\ \in D(G)$. *Then* $f$ *belongs to the centre* $Z(D(G))$ *if and only if the following conditions are met:*

(i) $C(x,a) = 0$ *unless* $xa = ax$; *and*

(ii) $C(x,a) = C(bxb^{-1}, bab^{-1})\ \forall\ b \in G$.

**Proof:** If $y, b, g, h \in G$ are fixed, we find on equating coefficients of $(g,h)$ in $f \cdot (y,b)$ and in $(y,b) \cdot f$ that $f \in Z(D(G))$ if and only if

$$\delta_{g,hb^{-1}ybh^{-1}}\ C(g, hb^{-1}) \;=\; \delta_{g,y}\ C(b^{-1}gb, b^{-1}h)\ \forall\ y, b, g, h \in G\ . \tag{4.2.8}$$

Suppose $f$ does belong to the center. Setting $y = g, x = b^{-1}gb, a = b^{-1}h$ in the above equation, we get

$$\delta_{bxb^{-1},baxa^{-1}b^{-1}}\ C(bxb^{-1}, bab^{-1}) \;=\; C(x,a)\ \forall\ x, a, b \in G\ ;$$

and conditions (i) and (ii) are seen to hold.

Conversely suppose conditions (i) and (ii) hold. We need to verify the condition in equation (4.2.8). For this note that if $g$ and $hb^{-1}$ do not commute, then ($b^{-1}gb$ and $b^{-1}h$ also do not commute and) condition (i) ensures that both $C(g, hb^{-1})$ and $C(b^{-1}gb, b^{-1}h)$ must vanish; while if $g$ and $hb^{-1}$ do commute, then an appeal to condition (ii) clinches the validity of equation (4.2.8), as desired. □

It will help to set up some notation: the symbol $\mathcal{C}(G)$ will denote the set of conjugacy classes of the group $G$, and $C$ will denote a typical element of $\mathcal{C}(G)$; we shall assume that we have chosen a representative $x(C) \in C$, and we set $H(C) = \{g \in G : gx(C) = x(C)g\}$. For $d = 0, 1, 2, \cdots$, we shall write

$$X_d(G) \;=\; \{(a_1, b_1, \cdots, a_d, b_d) \in G^{2d} : [a_1, b_1]\ \cdots\ [a_d, b_d] \;=\; 1\}\ .$$

COROLLARY 4.2.2

$$dim\ Z(D(G)) \;=\; \sum_{C \in \mathcal{C}(G)} |\mathcal{C}(H(C))|\ .$$

**Proof:** Lemma 4.2.1 is seen to imply that the dimension of $Z(D(G))$ is precisely the number of orbits in $X_1(G)$, under the

$G$-action given by $g \cdot (a, b) = (gag^{-1}, gbg^{-1})$; but a moment's thought shows that the collection of points $\{(x(C), g) : g \in \Lambda_C, C \in \mathcal{C}(G)\}$ - where $\Lambda_C$ is a set of representatives from distinct members of $\mathcal{C}(H(C))$ - contains exactly one element from each $G$-orbit in $X_1(G)$. □

PROPOSITION 4.2.3 *(a) There exists a bijection from* $\{(C, \pi) : \pi \in \widehat{H(C)}, C \in \mathcal{C}(G)\}$ *to the set of minimal central projections of* $D(G)$*, given by:* $e(C, \pi) = \frac{d_\pi |C|}{|G|} \sum_{x,a \in G} C(x, a)\ (x, a)$*, where it is required that:*

*(i)* $C(x, a) = 0$ *unless* $x \in C$ *and* $ax = xa$*; and*

*(ii)* $C(gx(C)g^{-1}, gag^{-1}) = \chi_\pi(a)\ \forall\ a \in H(C),\ g \in G$.

*(b) dim* $D(G)e(C, \pi) = (d_\pi |C|)^2$.

**Proof:** (a) It is straightforward to verify that the above prescription unambiguously defines an element $e(C, \pi)$ of $D(G)$ which in fact belongs to the centre in view of Lemma 4.2.1. If the convolution product of two functions on a finite group $H$ is defined by the rule

$$(\phi * \psi)(h) = \sum_{uv=h} \phi(u)\psi(v) ,$$

then we know that

$$\chi_\pi * \chi_\rho = \delta_{\pi,\rho} \frac{|H|}{d_\pi} \chi_\pi \ \forall\ \pi, \rho \in \hat{H} . \qquad (4.2.9)$$

It is not hard to see, with this equation applied to the group $H(C)$, that $\{e(C, \pi) : \pi \in \widehat{H(C)}\}$ is a collection of mutually orthogonal idempotents, for each $C \in \mathcal{C}(G)$. On the other hand, the definition of the product in $D(G)$ shows immediately that $e(C, \pi)e(D, \rho) = 0$ if $C \neq D \in \mathcal{C}(G)$.

$$\begin{aligned} |\{e(C, \pi) : \pi \in \widehat{H(C)}, C \in \mathcal{C}(G)\}| &= \sum_{C \in \mathcal{C}(G)} |\widehat{H(C)}| \\ &= \sum_{C \in \mathcal{C}(G)} |\mathcal{C}(H(C))| \\ &= dim\ Z(D(G)) \end{aligned}$$

by Lemma 4.2.1; hence we may conclude that the $e(C,\pi)$'s are necessarily minimal central projections of the $C^*$-algebra $D(G)$.

(b) We may deduce from the definitions that the coefficient of $(y,h)$ in $(y,h)\cdot e(C,\pi)$ is given by $\frac{d_\pi^2|C|}{|G|}\,1_C(y)$; hence the trace of right-multiplication by $e(C,\pi)$ is $(d_\pi\,|C|)^2$, and the proof is complete. □

LEMMA 4.2.4 *For any finite group* $H$, *and* $d = 0,1,\cdots$, *we have*

$$\frac{1}{|H|}\,|X_d(H)| \;=\; \sum_{\pi\in\hat{H}} \left(\frac{|H|}{d_\pi}\right)^{2d-2} .$$

**Proof:** Define $\theta_H^{(d)} : H \to \mathbb{Z}$ by

$$\theta_H^{(d)}(h) \;=\; |\{(a_1,b_1,\cdots,a_d,b_d)\in H^{2d} : [a_1,b_1]\,\cdots\,[a_d,b_d] \;=\; h\}|\,,$$

so that $|X_d(H)| \;=\; \theta_H^{(d)}(1)$. Hence we need to show that

$$\theta_H^{(d)}(1) \;=\; \sum_{\pi\in\hat{H}} \left(\frac{|H|^{2d-1}}{d_\pi^{2d-2}}\right) . \tag{4.2.10}$$

For this, observe first that $\theta_H^{(d)}$ is just the $d$-fold convolution power of $\theta_H^{(1)} \;=\; \theta$ (say). Then notice that $\theta$ is clearly constant on conjugacy classes, and is hence a linear combination of the $\chi_\pi$'s.

We assert that

$$\theta \;=\; \sum_{\pi\in\hat{H}} \frac{|H|}{d_\pi}\,\chi_\pi\,. \tag{4.2.11}$$

Fix $h \in H$, define $T_y : H \to H$ by $T_y(x) = hyxy^{-1}$ and regard $T_y$ as being extended linearly to a map of $\mathbb{C}H$ into itself; then

$$\theta(h) \;=\; \sum_{y\in H} \operatorname{Tr} T_y\,.$$

Let $\{e^\pi : \pi\in\hat{H}\}$ be the set of minimal central projections of $\mathbb{C}H$, and let $\{e^\pi_{ij} : 1\le i,j\le d_\pi\}$ be a set of matrix units for the (full matrix-) algebra $(\mathbb{C}H) \cdot e^\pi$; note that

$$\begin{aligned} T_y e^\pi_{ij} &= h y e^\pi_{ij} y^{-1} \\ &= \sum_{\rho,k,l,\mu,m,n,\lambda,p,q} (\rho_{kl}(h) e^\rho_{kl})(\mu_{mn}(y) e^\mu_{mn}) e^\pi_{ij} (\lambda_{pq}(y^{-1}) e^\lambda_{pq}) ; \end{aligned}$$

and hence the coefficient of $e^\pi_{ij}$ in $T_y e^\pi_{ij}$ is seen to simplify (in view of the mutual orthogonality of the $e^\rho$'s) to $\pi_{ii}(hy)\pi_{jj}(y^{-1})$.

Conclude that

$$\begin{aligned} \theta(h) &= \sum_{y \in H} Tr\ (T_y) \\ &= \sum_{y \in H} \sum_{\pi,i,j} \pi_{ii}(hy)\pi_{jj}(y^{-1}) \\ &= \sum_{y \in H} \sum_{\pi} \chi_\pi(hy)\chi_\pi(y^{-1}) \\ &= \sum_{\pi} (\chi_\pi * \chi_\pi)(h) \\ &= \sum_{\pi} \frac{|H|}{d_\pi} \chi_\pi(h) , \end{aligned}$$

thereby establishing equation (4.2.11).

An easy induction argument, coupled with equation (4.2.9), now shows that

$$\theta^{(d)}_H = (\theta * \overset{d\ times}{\cdots} * \theta) = \sum_{\pi \in \hat{H}} \left( \frac{|H|}{d_\pi} \right)^{2d-1} \chi_\pi .$$

We obtain the desired equation (4.2.10) by evaluating the above equation at 1. □

We conclude by fulfilling the promise made in the second paragraph of this section.

Proposition 4.2.5 *For* $d = 0, 1, 2, \cdots$, *we have:*

$$dim\ V_{sur}(\Sigma_d) = dim\ V_{tri}(\Sigma_d) .$$

**Proof:** In view of the familiar - see [Spa] - presentation of $\pi_1(\Sigma_d)$, we see from equation (4.1.4) that $dim\ V_{tri}(\Sigma_d)$ is nothing but the cardinality of the space $G\backslash X_d(G)$ of $G$-orbits in $X_d(G)$ (with respect to the 'diagonal conjugation action').

It is a consequence of Burnside's Lemma - see [Bur] - that

$$\begin{aligned} |G\backslash X_d(G)| &= \frac{1}{|G|}\sum_{g\in G} |\{x \in X_d(G) : g\cdot x = x\}| \\ &= \frac{1}{|G|}\sum_{g\in G} |X_d(\{a \in G : ag = ga\})| \\ &= \frac{1}{|G|}\sum_{C\in\mathcal{C}(G)}\sum_{g\in C} |X_d(\{a \in G : ag = ga\})| \\ &= \frac{1}{|G|}\sum_{C\in\mathcal{C}(G)} |C|\,|X_d(H(C))| \\ &= \sum_{C\in\mathcal{C}(G)} \frac{1}{|H(C)|}\,|X_d(H(C))|\,. \end{aligned} \tag{4.2.12}$$

On the other hand, we know from Proposition 4.2.3 and Verlinde's formula that

$$\begin{aligned} dim\ V_{tri}(\Sigma_d) &= \sum_{C\in\mathcal{C}(G)}\sum_{\pi\in\widehat{H(C)}} \frac{|G|^{2d-2}}{(|C|d_\pi)^{2d-2}} \\ &= \sum_{C\in\mathcal{C}(G)}\left(\sum_{\pi\in\widehat{H(C)}} \frac{|H(C)|^{2d-2}}{(d_\pi)^{2d-2}}\right). \end{aligned} \tag{4.2.13}$$

An appeal to Lemma 4.2.4 and a comparison of equations (4.2.12) and (4.2.13) completes the proof of the proposition. □

# Appendix A

## A.1 Some topological preliminaries

We shall not be too technical in this section; we will be happy with a vague and intuitive understanding of the topological background. The reader who desires more details could do no better than to consult [Mil]. We shall be dealing with manifolds; a *d-dimensional manifold* is a 'space which locally looks like $d$-dimensional Euclidean space'. Some examples of a 1-dimensional manifold are given in the Figure A.1.

The most familiar example of a (2-dimensional) manifold is the surface of the earth or its idealised version, the '2-sphere'; another favourite of mathematicians which they customarily call a '2-torus' is just the surface of a tube or doughnut.

Further illuminating examples (of 2-folds) are provided by the cylinder, and the Möbius band, which are obtained by identifying a pair of parallel boundary edges of a rectangle with the same and opposite orientations, respectively. A major distinction between the Möbius band and the other examples discussed so far is that the former is *not orientable.* A loose way of saying this is that it is not possible to consistently define a 'right-handed frame'. We shall be primarily concerned with orientable 3-manifolds; we shall content ourselves with saying here that given any two orthogonal directions at a point in the manifold, there is a well-defined prescription which singles out one of two opposite directions for a third orthogonal direction.

The manifolds discussed so far have been 'manifolds without boundary'. We shall also be concerned in the text with the

Figure A.1: Some 1-folds

so-called '$d$-dimensional manifolds with boundary'. These are spaces whose points are of two kinds: the 'internal points' whose neighbourhoods look like the interior of the ball $\mathbb{B}^d$ in $\mathbb{R}^d$; and the 'boundary points' whose neighbourhoods look like the set $\{(x_1, \cdots, x_d) \in \mathbb{B}^d : x_d \geq 0\}$. (For example, the set $\{x \in \mathbb{R}^d : ||x|| \leq 1\}$ is a $d$-dimensional manifold with boundary; its boundary (= the set of its boundary points) is the $(d-1)$-sphere $S^{d-1} = \{x \in \mathbb{R}^d : ||x|| = 1\}$. It is true, in general, that the boundary of a $d$-dimensional manifold with boundary is a $(d-1)$-dimensional manifold without boundary. We can, and will, talk about manifolds with boundary which are orientable.

Throughout the text, when we say manifold, we mean a manifold with (possibly empty) boundary; and if $M$ is a manifold, the symbol $\partial M$ denotes the 'boundary of $M$'.

One of the central problems of topology is to determine when two manifolds are 'the same' (or homeomorphic to one another) or 'deformable to one another'. TQFTs, as discussed here, are a possible way of deciding if the answer to the second question above is negative; in other words, they yield 'invariants' which, by definition, have the property that 'equivalent objects yield the same invariants'.

## A.2 Simplicial complexes

We shall briefly recall the basic facts concerning simplicial complexes here. (We shall only deal with 'finite' simplicial complexes.)

DEFINITION A.2.1 *A (finite)* **simplicial complex** *is a collection* $K$ *(whose typical member will be denoted by* $s$*, and referred to as a simplex of* $K$*) of non-empty finite subsets of a set* $V$ *(whose typical member will be denoted by* $v$ *and called a vertex or 0-simplex of* $K$*) with the following properties:*

*(a) every singleton subset of* $V$ *is a simplex; and*

*(b) every non-empty subset of a simplex is a simplex.*

A simplex with exactly $d+1$ vertices is called a *d-simplex*. If the vertices of a $p$-simplex, say $s$, form a subset of the set of vertices of a $q$-simplex, say $s'$, then $s$ is called a *face* of $s'$.

To make contact with more familiar topological objects, it would probably help if we defined *the geometric* (also called *the standard) d-simplex* to be the set

$$\Delta^d = \{(x_0, x_1, \cdots, x_d) \in \mathbb{R}^{d+1} : x_i \geq 0 \ \forall i \ , \ \sum_{i=0}^{d} x_i = 1\} \ .$$

Thus, for $d = 0, 1, 2, 3$, the set $\Delta^d$ is nothing but a point, a closed line-segment, a closed triangle and its interior, a closed tetrahedron with its interior, etc.

The association of a 'geometric $d$-simplex' to an 'abstract $d$-simplex' can be generalised to a mapping - called *geometric realisation* - from the class of finite simplicial complexes to a class of compact subspaces of Euclidean spaces (usually termed polyhedra). The idea is that one takes a collection of 'geometric simplices', one for every 'abstract simplex' (of the same 'corresponding dimension'), and 'makes identifications as indicated by the simplicial complex'.This correspondence is intuitively quite clear, while the precise formulation might be slightly less palatable; so we content ourselves with a few cxamples.

(i) Let $K$ consist of the simplices

$$\{1,2,3\} \ , \ \{1,2,4\} \ , \ \{1',2',3'\} \ , \ \{1',2',4'\} \ , \ \{1,1'\}$$

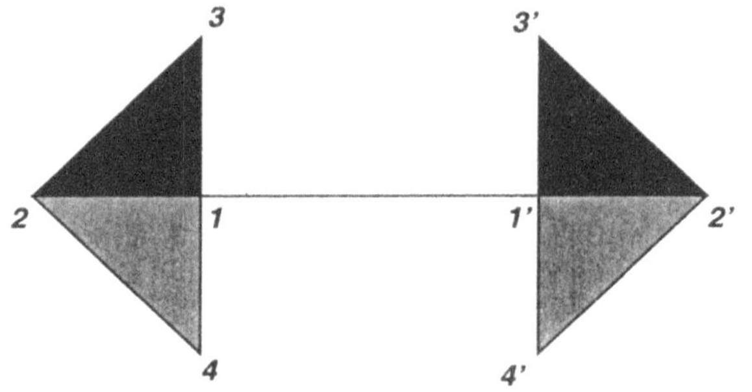


Figure A.2: Example 1

along with all the 'faces' of all the above simplices. Then the geometric realisation of $K$ is as in Figure A.2

(ii) If $K$ consists of the collection of all non-empty subsets of a set with $(d+1)$ elements, then the geometric realisation $|K|$ is nothing but what we have called the geometric $d$-simplex.

On the other hand, if $K$ consists of all non-empty *proper* subsets of a set with $(d+1)$ elements, then the geometric realisation $|K|$ is the boundary $\partial\Delta^d$ of the standard $d$-simplex.

In general, if $K$ is a simplicial complex, and $d$ is fixed, then the collection of all the $q$-simplices of $K$ for $q \leq d$ is also a simplicial complex, which is called the $d$-skeleton of $K$.

It is a fact - see [Moi] - that every compact 3-manifold (with or without boundary) 'admits a triangulation', meaning that it is homeomorphic to the geometric realisation of some (finite) simplicial complex.

It will help to recall some facts, at this stage, about the so-called *edge-path-groupoid of a simplicial complex.* So let $K$ be a simplicial complex. (In particular, we could take $K$ to be the triangulation $\Delta$ of a manifold.)

By an 'edge' of $K$ will be meant an ordered pair $e = (v_0, v_1)$ of (not necessarily distinct) 0-simplices of $K$ which belong to some simplex of $K$ (in fact a $q$-simplex, where $q$ may be chosen to be equal to 0 or 1). For the above edge $e$, we shall refer to $v_0$ as the *source* of $e$, and to $v_1$ as the *range* of $e$.

By an edge-path $\gamma$ in $K$, we shall mean a string $e_1 \cdot e_2 \cdots e_n$, with $n \geq 1$, such that:

(i) each $e_i$ is an edge of $K$;

(ii) for each $1 \leq k < n$, the source of $e_{k+1}$ and the range of $e_k$ agree.

The source and range of a path $\gamma$, as above, are defined to be the source of $e_1$ and the range of $e_n$, respectively.

Two edge-paths, say $\gamma$ and $\gamma'$, are considered to be equivalent if there exists a finite chain $\gamma = \gamma_0, \gamma_2, \cdots, \gamma_m = \gamma'$ of edge-paths such that for each $1 \leq t \leq m$, there exists a set $T = \{u, v, w\}$ of 0-simplices of $K$ satisfying: (i) $T$ is a simplex of $K$ (in fact, a $q$-simplex, where $q \leq 2$); and (ii) one of the edge-paths $\gamma_{t-1}$ or $\gamma_t$ contains the edge $(u, w)$, while the other is obtained by making the single change of replacing the edge $(u, w)$ by $(u, v) \cdot (v, w)$.

The definitions show that equivalent edge-paths have the same source and range; i.e., the notions of source and range make sense of equivalence classes of edge-paths. There is also a natural (concatenation-) product of two equivalence classes $[\gamma_1]$ and $[\gamma_2]$ which satisfy the compatibility condition $s(\gamma_2) = r(\gamma_1)$. It is painless to see that the collection of equivalence classes of edge-paths acquires the structure of a groupoid with respect to these operations. This is called the *edge-path groupoid* of the simplex $K$, which we shall denote by the symbol $EPG(K)$.

It is true - see [Spa] - that any path - say $\kappa$ - in (the geometric realisation $|K|$) of $K$, with end-points being 0-simplices, is homotopic rel. end-points to an 'edge-path' - say $\gamma$; and further, if $\kappa'$ and $\gamma'$ are also so related, then $\kappa$ and $\kappa'$ are homotopic rel. end-points if and only if $\gamma$ and $\gamma'$ are equivalent as edge-paths.

In other words, there is a natural identification of the edge-path groupoid of $K$ and the fundamental groupoid of $|K|$.

# Bibliography


[Ati] Atiyah, M., *Topological quantum field theory*, Publ. Math. IHES, **68**, (1989), 175-186.

[BHMV] Blanchet, C., Habegger, N., Masbaum, G., and Vogel, P., *Topological quantum field theories derived from the Kaufmann bracket*, Topology, **34**, (1995), 883-927.

[Bur] Burnside, W., *Theory of groups of finite order*, Second Edition, Cambridge University Press, Cambridge 1911.

[DPR1] Dijkgraaf, R., Pasquier, V., and Roche, Ph., *Quasi Hopf algebras, group cohomology and orbifold models*, Nuclear Physics B(Proc. Suppl.), **18**, (1990), 60-72.

[DPR2] Dijkgraaf, R., Pasquier, V., and Roche, Ph., *Quasi-quantum groups related to orbifold models*, Proceedings of the International Colloquium on Modern Quantum Field Theory, World Scientific, Singapore, 375-383, 1991.

[DVVV] Dijkgraaf, R., Vafa, C., Verlinde, E., and Verlinde, H., *The operator algebra of orbifold models*, Communications in Mathematical Physics, **123**, (1989), 485-526.

[DW] Dijkgraaf, R., and Witten, E., *Topological gauge theories and group cohomology*, Communications in Mathematical Physics, **129**, (1990), 393-429.

[EK] Evans, D.E., and Kawahigashi, Y., *Quantum symmetry on operator algebras*, Oxford University Press, 1998.

[JS] Jones, V., and Sunder, V.S., *Introduction to subfactors*, London Math. Soc. Lecture Note Series, 234, Cambridge University Press, Cambridge, 1997.

[Mil] Milnor, J.W., *Topology from a differentiable viewpoint*, Univ. of Virginia Press, Charlottesville, VA, 1965.

[Moi] Moise, E.E., *Affine structures in 3-manifolds. VIII. Invariance of the knot-types; local tame embedding*, Ann. Math., (2) **59**, (1954), 159-170.

[Oc1] A. Ocneanu, *Quantized groups, string algebras and Galois theory for algebras*, Operator Algebras and Appl., Vol. 2 (Warwick 1987), London Math. Soc. Lecture Notes Ser. Vol. 136, pp. 119-172, Cambridge University Press, 1988.

[Oc2] A. Ocneanu (Lecture Notes written by Y. Kawahigashi), *Quantum symmetry, differential geometry of finite graphs, and classification of subfactors*, Univ. of Tokyo Seminar Notes, 1990.

[Pac] Pachner, U., *Bistellare Aquivalenz kombinatorischer Mannigfaltigkeiten*, Arch. Math., **30**, 1978, 89-91.

[Spa] Spanier, E.H., *Algebraic topology*, McGraw-Hill, 1966.

[Sun] Sunder, V.S., *$II_1$ factors, their bimodules and hypergroups*, Trans. of AMS, **330**, (1992), 227-256.

[Tur] Turaev, V.G., *Quantum invariants of knots and 3-manifolds*, de Gruyter Studies in Math., vol. 18, 1994.

[TV] Turaev, V., and Viro, O.Y., *State sum invariants of 3-manifolds and quantum 6j-symbols*, Topology, **31**, 1992, 865-902.

## Bibliographical Remarks

The presentation in §1.2 is a leisurely expanded version of the first few pages of the paper [BHMV]. One reason for adopting this approach, rather than merely stating the axioms for a TQFT as in [Ati], is to draw attention to the interesting problem of whether the TQFT arising from a finite-depth subfactor, as in Chapter 3, is cobordism generated.

The treatment, in §1.3, of fusion algebras is modelled after that of the paper [Sun], where the term 'hypergroup' was used for what is called the 'distinguished basis' of a finite-dimensional fusion algebra.

The approach in Chapter 2 is essentially equivalent to the one in [EK], the justification for the difference here being that the device, adopted here, of introducing an extraneous 'total order' and then establishing that the invariant is independent of that order eliminates much of the possible confusion that results from consistently ignoring the orientations on the edges.

We have adopted the phrase *(2-3 condition)* in place of what is referred to, in [EK], as the *pentagonal relation*, since the latter phrase has a very definite meaning in the context of monoidal categories, and it was not entirely clear to us that that condition is 'the same' as our condition here.

We have stated Pachner's result in a form that is convenient for our use. (It is easily seen to be equivalent to the formulation in [Tur].) The original result is for closed manifolds, and the extension to 3-manifolds with boundary was obtained in [TV], Appendix 1. Further, the proof we have presented of the fact that the invariant is indeed independent of the triangulation is a convex combination of the ones found in [EK] and [Tur].

As for the unproved facts in Chapter 3, either their proofs or information as to the location of such proofs may be found in [JS]. The material in §3.5 is essentially from [EK] - although we have coined some of the terminology and notation used in §3.5, in the hope that it may clarify exposition; in particular, Evans and Kawahigashi are entirely blameless for our discussion of 'tethered marked courses' and other such objects.

Most of the facts in Chapter 4 can be found - albeit in language that physicists are usually more comfortable with than mathematicians - in [DW], [DVVV], [DPR1] and [DPR2]. (Also, the statement of Proposition 4.1.2 is stated as an exercise in [EK].) The formulation given in §4.2 of Verlinde's formula is found in [Tur].

The discussion of simplicial complexes and the edge-path groupoid in the Appendix is modelled after [Spa].

# Index

For Product Safety Concerns and Information please contact our
EU representative GPSR@taylorandfrancis.com Taylor & Francis
Verlag GmbH, Kaufingerstraße 24, 80331 München, Germany